ELECTROMECHANICS OF PARTICLES

ELECTROMECHANICS OF PARTICLES

THOMAS B. JONES

University of Rochester

CAMBRIDGE
UNIVERSITY PRESS

CAMBRIDGE UNIVERSITY PRESS
Cambridge, New York, Melbourne, Madrid, Cape Town, Singapore, São Paulo

Cambridge University Press
The Edinburgh Building, Cambridge CB2 2RU, UK

Published in the United States of America by Cambridge University Press, New York

www.cambridge.org
Information on this title: www.cambridge.org/9780521431965

© Cambridge University Press 1995

First published 1995
This digitally printed first paperback version 2005

A catalogue record for this publication is available from the British Library

Library of Congress Cataloguing in Publication data
Jones, T. B. (Thomas B.)
Electromechanics of Particles / Thomas B. Jones.
p.   cm.
Includes bibliographical references (p.).
ISBN 0–521–43196–4
1. Dielectrophoresis.   2. Particles.   3. Electromechanical analogies.   I. Title.
QC585.7.D5J6   1995
543′.0871 – dc20   94-38849
                                                                          CIP

ISBN-13 978-0-521-43196-5 hardback
ISBN-10 0-521-43196-4 hardback

ISBN-13 978-0-521-01910-1 paperback
ISBN-10 0-521-01910-9 paperback

To my wife, Mary, and to my daughters, Laura and Audrey

Contents

Preface   page xiii
Nomenclature   xvii

1 Introduction   1
  1.1 Background and motivation   1
  1.2 Objectives of this book   2
  1.3 Limitations and caveats   3
2. Fundamentals   5
  2.1 Introduction   5
    A. Electromechanics of particles   5
    B. Force on an infinitesimal dipole   6
    C. Torque on a dipole   8
  2.2 Lossless dielectric particle in an electric field   9
    A. Induced multipolar moments   9
    B. Effective dipole moment of dielectric sphere in
       dielectric medium   10
    C. Conducting sphere in uniform DC field   12
    D. Lossless spherical shell in lossless medium with
       uniform field   12
    E. Lossless dielectric sphere in lossless dielectric medium
       in field of point charge   14
  2.3 Dielectric particle with loss in an electric field   16
    A. Homogeneous sphere in an AC electric field   17
    B. Shells with ohmic loss in an AC electric field   20
    C. Summary   21
  2.4 Effective moment calculation of force and torque   24
    A. Hypothesis and definitions   24
    B. Lossless particles   25
    C. Particles with ohmic (or dielectric) loss   28
  2.5 Theory of Sauer and Schlögl   30

	A.	Summary of their result	31
	B.	Reconciliation with effective moment method	31
	C.	Discussion	32

3 Dielectrophoresis and magnetophoresis   **34**

3.1 Introduction   34
   A. Phenomenological definition   35
   B. Further delineation of dielectrophoretic effect   35
3.2 DEP phenomenology for lossless spherical particle   36
   A. Dielectrophoretic force expression   36
   B. Phenomenology of DEP   37
   C. Higher-order multipolar force terms   38
3.3 Frequency-dependent DEP phenomenology   39
   A. Homogeneous sphere with ohmic loss   39
   B. Time-dependent DEP   42
   C. Layered spherical shells with loss   42
   D. Generalized definition for +DEP and −DEP   43
   E. Examples of lossy spherical shells   43
3.4 DEP levitation   49
   A. DEP levitation theory   49
   B. Dynamic model of DEP levitator   52
   C. Passive DEP levitation   55
   D. Feedback-controlled DEP levitation (+DEP)   58
   E. Discussion of DEP levitation technique   60
3.5 Magnetophoresis   62
   A. Theory   63
   B. Magnetically linear particle   65
   C. Nonlinear magnetic media   66
   D. Magnetophoresis with eddy current induction   69
   E. Force on superconducting particles   73
3.6 Applications of dielectrophoresis and magnetophoresis   74
   A. Biological DEP   74
   B. Microactuators   78
   C. DEP mineralogical separations   79
   D. High-voltage liquid insulation   79
   E. MAP separation technologies   80
   F. Magnetic particle levitation   81

4 Particle rotation   **83**

4.1 Introduction   83
4.2 Theory for particle rotation   84
   A. Particle interactions with a rotating electric field   84

| | | B. | Electric torque on homogeneous dielectric particle with ohmic loss | 86 |

		B.	Electric torque on homogeneous dielectric particle with ohmic loss	86
		C.	Turcu's bifurcation theory	88
	4.3		Rotational (relaxation) spectra	92
		A.	General theory	92
		B.	Sample spectra	94
		C.	Argand diagrams	98
	4.4		Quincke rotation in DC electric field	98
		A.	Theory	98
		B.	Experimental results	100
	4.5		Rotation of magnetizable particles	101
		A.	Induction	101
		B.	Nonlinear effects	103
	4.6		Applications of electrorotation	105
		A.	Cell characterization studies	106
		B.	Cell separation	108
		C.	Practical implications of the Quincke effect	108
5	**Orientation of nonspherical particles**			**110**
	5.1		Introduction	110
	5.2		Orientation for lossless homogeneous ellipsoids	111
		A.	Isotropic ellipsoid in a uniform electric field	111
		B.	Alignment torque expressions	113
		C.	Two special cases: prolate and oblate spheroids	115
		D.	DEP force on ellipsoidal particle	118
	5.3		Orientation for lossy dielectric ellipsoids	119
		A.	Theory for homogeneous particles	119
		B.	Alignment behavior of homogeneous particles	120
		C.	Orientation of layered particles	121
	5.4		Experimental orientational spectra	124
	5.5		Static torque on suspended particle	125
		A.	Basic model for torque calculation	126
		B.	Rotational torque on suspended particle	128
		C.	Alignment torque on suspended particle	130
	5.6		Orientation of magnetizable particles	132
		A.	Magnetically linear particles with anisotropy	132
		B.	Prolate spheroidal crystals	134
		C.	Thin disks and laminae	135
		D.	Isotropic particle with remanent magnetization	135
	5.7		Applications of orientational phenomena	136
		A.	Dielectric particles	136

      B.   Magnetizable particle applications   137

6   Theory of particle chains   139
  6.1   Introduction to chaining and review of previous work   139
  6.2   Linear chains of conducting spheres   142
      A.   Solution using the method of images   142
      B.   Solution using geometric inversion   149
      C.   Alignment torque for chain of two identical conducting
          spheres   153
      D.   Spheres of unequal radii   154
      E.   Intersecting spheres   155
      F.   Discussion of results for short chains of
          contacting particles   156
  6.3   Chains of dielectric (and magnetic) spheres   159
      A.   Simple dipole approximation   160
      B.   Accuracy considerations   162
      C.   General expansion of linear multipoles   164
      D.   Experimental measurements on chains   166
  6.4   Frequency-dependent orientation of chains   172
  6.5   Heterogeneous mixtures containing particle chains   173
      A.   Mixture theory   175
      B.   Suspensions of chains   177
  6.6   Conclusion   180

7   Force interactions between particles   181
  7.1   Introduction   181
  7.2   Theory   181
      A.   Force between conducting spheres   183
      B.   Force between dielectric spheres   185
      C.   Current-controlled interparticle forces   187
  7.3   Experiments   188
      A.   Interparticle force measurements   188
      B.   Measurements on longer chains and planar arrays   189
      C.   Nonlinear effects   192
  7.4   Electrostatic contributions to adhesion   194
      A.   Phenomenological force expression   194
      B.   Image force contributions   196
      C.   Detachment force contribution   203
      D.   Induced moment contributions   204
      E.   Generalized model of Fowlkes and Robinson   204
  7.5   Mechanics of chains and layers   207
      A.   Chains of magnetizable particles   209

B.   Mechanics of particle beds   209
7.6 Discussion of applications   211
    A.   Electrofusion of biological cells   211
    B.   Chaining in electrorheological fluids   212
    C.   Electrofluidized and electropacked beds   214
    D.   Magnetopacked and magnetofluidized beds   215
7.7 Closing prospect   216

Appendix A: *Analogies between electrostatic, conduction, and*
              magnetostatic problems   218
Appendix B: *Review of linear multipoles*   222
Appendix C: *Models for layered spherical particles*   227
Appendix D: *Transient response of ohmic dielectric sphere to a*
              suddenly applied DC electric field   236
Appendix E: *Relationship of DEP and ROT spectra*   238
Appendix F: *General multipolar theory*   248
Appendix G: *Induced effective moment of dielectric ellipsoid*   251
References   253
Index   263

Preface

As a consequence of their electrical and/or magnetic properties, all particles experience forces and torques when subjected to electric and/or magnetic fields. Furthermore, when they are electrically charged, polarized, or magnetized, closely spaced particles often exhibit strong mutual interactions. In this book, I focus on these particle–field interactions, referred to collectively as *particle electromechanics*, by delineating common phenomenology and by developing simple yet general models useful in predicting electrically and magnetically coupled mechanics. The objective is to bring together diverse examples of field–particle interactions from many technologies and to provide a common framework for understanding the relevant electromechanical phenomena. It may disappoint some readers to learn that, despite the rather general definition offered for particle electromechanics, I restrict attention to particles in the size range from approximately 1 micron (10^{-6} m) to 1 millimeter (10^{-3} m). Though many of the ideas developed here indeed carry over into the domain of ultrafine particles, the lower limit recognizes that other phenomena, such as van der Waals forces and thermal (Brownian) motion, become important below one micron. The upper limit is consistent with a reasonable definition for a classical particle.

Chapter 1 introduces the subject, provides a definition for particle electromechanics, and adds some caveats to inform the reader of the book's limitations. Chapter 2 unveils the fundamental *effective moment* concept employed throughout the book in the calculation of electromechanical forces and torques. It also uses multipolar expansion methods to solve for the induced moments for a particle experiencing a strongly nonuniform field and exploits the analogy between electrostatic and magnetostatic problems to reveal how the results for a dielectric particle can be applied to a magnetizable particle. Chapter 3 harnesses the effective moment method to derive dielectrophoretic (DEP) and magnetophoretic (MAP) force expressions for small particles in nonuniform AC electric and magnetic fields, respectively. The frequency-dependent DEP force on dielectric particles with ohmic or dielectric loss is examined with reference to dielectrophoretic levitation. Particle rotation is the subject of Chapter 4, where the effective moment method aids us in calculating the frequency-dependent torque on a particle due to a rotating electric field. Spontaneous Quincke rotation (in a static, DC electric field) is considered as a special case. The behavior of magnetizable and electrically conductive particles in a rotating magnetic field is examined by drawing on the close analogy to rotating machines. Chapter 5 concerns nonspherical (ellipsoidal) particles in uniform fields, principally the frequency-dependent orientation of lossy dielectric ellipsoids in AC electric fields and the alignment of magnetic crystals in a DC magnetic field.

The remainder of the book focuses on the mutual interactions between closely spaced particles. Chapter 6 describes methods for determining the effective multipolar moments of linear particle chains. These effective moments provide the means to calculate net DEP forces and electrical torques exerted upon chains by electric or magnetic fields. Several different approaches useful in modeling the particle interactions are covered, including the *method of images* and *method of inversion*, both applicable only to conductive particles, and a straightforward *multipolar expansion* technique, applicable to dielectric and magnetizable particles. This chapter also examines the effect of nonlinear magnetization upon the interactions of ferromagnetic particle chains and introduces a simple model for heterogeneous mixtures of particles agglomerated into short chains.

Chapter 7 exploits the particle interaction models of the previous chapter to determine the mutual forces of attraction between closely spaced particles. Experimental force measurements obtained with magnetizable spheres are compared to predictions of a linear multipolar expansion. One section of this chapter, devoted to the electrostatic contributions to surface adhesion, highlights practical methods for estimating these forces in realistic situations. An extensive review of some particulate technologies where chaining is important concludes this chapter.

A nomenclature section lists all important algebraic symbols used in the mathematical expressions. All literature references, cited by author and date of publication in the text, are listed alphabetically at the end of the book. A large set of appendixes at the end of the book covers the details of certain mathematical derivations and results. The choice of SI units for all variables and mathematical expressions throughout the text reflects the significant interdisciplinary character of particle electromechanics as well as the personal choice of the author.

This book evolved from lecture notes used by the author in a graduate course entitled "Particle Electromechanics" first offered at the University of Rochester in 1985. Among many who helped bring it into being, I am privileged to mention Dr. William Y. Fowlkes, Dr. Kelly S. Robinson, and Dr. Bijay Saha of Eastman Kodak Company; Dr. Ruth D. Miller of Kansas State University; John Kraybill of Weston Controls; Prof. Herman P. Schwan and Prof. Kenneth R. Foster of the University of Pennsylvania; Prof. Edwin L. Carstensen of the University of Rochester; and Dr. P. Keith Watson and Dr. Robert Meyer of Xerox Corporation. I have benefited particularly from extensive technical collaboration and interchanges with Professor Karan V. I. S. Kaler of the University of Calgary and Dr. Masao Washizu of Seikei University (Tokyo). Special thanks go to Dr. Friedrich A. Sauer of the Max Planck Institüt für Biophysik in Frankfurt, Germany, who introduced me to his seminal work on the fundamental electrodynamical formulation of dielectrophoresis in lossy media during the summer of 1985. Dr. Michael P. Perry of the General Electric Company provided valuable advice in managing the daunting task of technical book writing. Finally, a special debt of gratitude is due to the late Prof. James R. Melcher of MIT, who nurtured my enthusiasm for electromechanics and taught me much of what I know about the subject.

This book is based in part upon research supported by a series of grants from the Particulate and Multiphase Processes Program of the National Science Foundation, the Copy Products Research and Development Division of Eastman Kodak Company, and the Hitachi Research Laboratories. I am deeply grateful for this support and also for two travel grants from the North Atlantic Treaty Organization. Finally, a sabbatical leave granted by the University of Rochester during the fall of 1990 provided the opportunity to initiate the task of writing.

Having benefited from those who have encouraged, taught, or studied with me, I can lay full claim only to the errors found in this book. Concerning these errors, readers are kindly requested to bring them to my attention.

Nomenclature

A	area of charge patch, Equation (7.14)
$\overline{A}$	magnetic vector potential, Equation (3.61)
A, B	coefficients of dipole solution for potential used in Equations (2.8a,b)
A_n, B_n	coefficients of multipolar potential terms in Equations (2.17a,b)
A, B, C, D	coefficients of dipole solution for potential of layered sphere in uniform field (Appendix C)
A_n, B_n, C_n, D_n	coefficients of multipolar terms for potential of layered sphere in field of point charge (Appendix C)
$\overline{B}$	magnetic flux density
D	spacing of source charges $\pm Q$ shown in Figure 6.2
$\overline{D}$	electric displacement vector
$\underline{D}(\omega)$	complex quantity for conducting magnetizable sphere in time-varying field defined by Equation (3.64)
$\overline{E}$	electric field vector
E_{gap}	electric field in gap between closely spaced particles
E_{max}	breakdown-limited electric field between closely spaced particles
E_0	externally applied electric field
$E_{\parallel}, E_{\perp}$	parallel and perpendicular components of electric field
$\overline{F}$	force vector
$F_{m,n}$	attractive force between two aligned linear multipoles of orders m and n in Equation (7.1)
G	feedback gain coefficient in Equation (3.43)
$\overline{H}$	magnetic intensity vector
H_c	coercive magnetic field value
H_{gap}	magnetic field in gap between closely spaced particles
I	moment of inertia of spherical particle used in Equation (4.23)
$I_{\pm n/2}$	modified half-integer order Bessel functions of first kind

$\bar{J}$	electric current density vector
K	Clausius–Mossotti functions as defined by Equation (2.13)
$K^{(n)}$	polarization coefficient of nth order linear multipole defined by Equation (2.23)
K_0, K_∞	low- and high-frequency limits of complex Clausius-Mossotti function
K_x, K_y, K_z	polarization factors along three principal axes of ellipsoidal particle, Equation (5.27)
$\bar{K}$	surface electric current density (Appendix C)
$\underline{L}(\omega) = R^3\underline{D}/2$	magnetization coefficient for conductive magnetizable sphere, Equation (3.63)
L_e	characteristic length of levitation electrodes
L_x, L_y, L_z	depolarization factors along x, y, and z axes for ellipsoid defined by Equation (5.4)
$L_\parallel, L_\perp$	depolarization factors parallel and perpendicular to long axis of prolate spheroidal particle, Equation (5.13)
$\bar{M}$	magnetization vector
M_n	normal component of magnetization vector
M_{rem}	remanent magnetization per unit volume
M_{sat}	saturation magnetization per unit volume
N	integer, number of spheres in linear chain
N_p	number of particles distributed in sphere of radius R_0
P	levitation equilibrium point: (x_0, y_0, z_0)
$\bar{P}$	polarization vector
$P_n(\cos\theta)$	Legendre polynomial of order n
Q	source charge for method of images
R	radius of spherical particle
R_α	radii of two identical intersecting spheres shown in Figure 6.10
R_a, R_b	radii of two interacting spherical particles shown in Figure 6.8b
$R_m = R\sqrt{\omega\mu_2\sigma_2/2}$	magnetic Reynolds number of conductive spherical particle used in Equation (3.68)
R_0	radius of equivalent sphere of homogeneous permittivity ε_{eff} defined in Figure 6.22
S_1, S_2	functions defined by Equations (7.19a,b)
T	temperature
$\bar{T}^e$	vector torque of electrical origin
T^η	viscous drag torque on rotating particle defined by Equation (4.7)
$\bar{T}^m$	vector torque of magnetic origin
V	voltage, or particle volume

V_0	equilibrium voltage for levitated particle at point P defined by Equation (3.34c)
X, Y	coefficients of magnetostatic potential solution in Equations (3.46a,b)
Z_1, Z_2	coefficients defined by Equations (G.1) and (G.2)
$a = R_1/R_2$	ratio of radii of spherical shell in Figures 2.3 and 3.4a
a, b, c	semimajor axes of ellipsoidal particle shown in Figure 5.1
b	damping coefficient in Equations (3.33a,b)
c	constant used in Equation (6.59)
c_m	cell membrane capacitance per unit area in Figures 3.3a and 3.4a
c_x, c_y, c_z	direction cosines
$\bar{d}$	vector distance between charges in finite dipole
d_n	distance of nth image charge from midplane of particle chain
e	eccentricity of spheroidal particle used in Equation (5.15)
$f = q_p/q$	patch fraction defined in Section 7.4B
$f(N), f(\theta)$	distribution functions for chain mixtures used in Section 6.5B
$\bar{g}$	acceleration due to gravity (9.81 m/s^2)
g_m	transmembrane conductance
i, j	integer indices
$j = \sqrt{-1}$	square root of minus one
k	Boltzmann's constant (1.38 • 10^{-23} J/deg K)
$\bar{m}$	magnetic dipole moment vector
m, n	integer indices
m_b	buoyant mass of particle in fluid
m_{eff}	effective magnetic dipole moment
m_{perm}	permanent moment of magnetized particle in air, Section 5.6D
m_{rem}	remanent moment of magnetized particle in Equation 4.37
$\hat{n}$	unit normal to surface
$\bar{p}$	electric dipole moment vector
$p^{(n)}$	moment of nth order linear multipole
p_{eff}	effective electric dipole moment
$\underline{p}_{eff} = p_{eff}/p_0$	normalized effective electric dipole moment
P_N	dipole moment of linear chain of N perfectly conducting spheres
$p_0 = 4\pi\varepsilon_1 R^3 E_0$	dipole moment of perfectly conducting sphere

$p_\parallel, p_\perp$	components of dipole moment parallel and perpendicular to long axis of prolate spheroid or chain
$\underline{p}_\parallel, \underline{p}_\perp$	parallel and perpendicular effective electric dipole moments normalized to p_0
q	electric charge
q_c	charge at center of dielectric particle, Figure 7.15
q_n	magnitude of nth image charge within conducting particle chain
q_p	charge on dielectric particle, Figure 7.15
$\vec{r}$	position vector
s	complex frequency variable used in Equation (E.1); variable of integration in Section 5.2
$s_\alpha, s_\beta, \dots$	poles of $K(s)$ defined in Appendix E
t	time
$u(t)$	unit step function
$v'(t)$	feedback voltage defined by Equation (3.34c)
x	variable of integration in Kramers–Krönig relations
$\hat{x}, \hat{y}, \hat{z}$	unit vectors in Cartesian coordinate system
x_1, x_2, x_3	Cartesian coordinates (x, y, z) in indicial notation (Section 3.4A)
Δ	uniform thickness of layer on spherical particle, Appendix C
$\Delta_x, \Delta_y, \Delta_z$	thicknesses of confocal layers in ellipsoid shown in Figure 5.4a
Ξ	center-to-center spacing of interacting spheres in Figure 7.2
$\Sigma_1, \Sigma_2, \Sigma_3$	summations defined by Equations (6.28a,b,c)
Σ	closed surface of integration
Φ	electrostatic potential function
Ψ	electromechanical potential function in Equation (3.24)
Ψ_1, Ψ_2	magnetostatic potential functions defined by Equations (3.46a,b)
Ω	angular velocity of particle
Ω_{eq}	equilibrium rotational speed of particle defined by Equation (4.6)
Ω_f	terminal angular velocity of Quincke rotation defined by Equation (4.27)
α	lagging angle between rotating field and dipole moment shown in Figure 4.2; angle of intersection for spheres in Figure 6.10
α, β, γ	indices representing right-hand, ordered sequence of Cartesian coordinates: $x, y, z, x, \dots$

$\alpha_1, \alpha_2, \cdots$	coefficients of field expansion along axis of axisymmetric field from Equation (3.32)
$\beta = R_b/R_a$	ratio of radii of chained spheres shown in Figure 6.8a
$\gamma = a/b$	ratio of axes of spheroid shown in Figure 5.2a
$\delta = \Xi - 2R$	gap spacing between two interacting particles
ε	dielectric permittivity
$\bar{\bar{\varepsilon}}$	tensor permittivity of anisotropic particle
$\underline{\varepsilon} = \varepsilon + \sigma/j\omega$	complex permittivity
ε'	real part of complex permittivity
ε''	negative imaginary part of permittivity
ε_2'	effective permittivity of spherical shell, Equation (C.4)
$\underline{\varepsilon}_\Sigma$	complex surface permittivity used in Equation (C.14)
ε_s	permittivity of dielectric substrate in Figure 7.12
ε_0	permittivity of free space ($8.854 \cdot 10^{-12}$ F/m)
ε_c, σ_c	permittivity and conductivity of cell cytoplasm, Section 3.3E
ε_w, σ_w	permittivity and conductivity of cell wall, Section 3.3E
ζ	distance of charge q from center of sphere in Figure 2.4
$\zeta(n)$	Riemann-zeta function of integer argument n
η_1	dynamic viscosity of fluid medium
θ	polar angle in spherical coordinates
$\vartheta_\|, \vartheta_\perp$	polarization coefficients for chains of unequal conductive particles in Equations (6.27a,b)
$\kappa = \varepsilon/\varepsilon_0$	dielectric constant
κ_s	dielectric constant of substrate in Figure 7.15
$\underline{\kappa} = \kappa' - j\kappa''$	complex dielectric constant
λ	exponent of electric field defined in Section 7.2C
λ_n	coefficients of neutralizing charges, Equation (6.12)
μ	magnetic permeability
μ_0	permeability of free space ($4\pi \cdot 10^{-7}$ H/m)
$\mu_R = \mu_2/\mu_1$	relative permeability
$\xi = (\kappa_s - \kappa_1)/(\kappa_s + \kappa_1)$	coefficient of images in dielectric plane used in Section 7.4B
ξ_n	distance of neutralizing charges from midplane in Figure 6.4c
ρ	radial coordinate in cylindrical coordinates
ρ_1, ρ_2	mass densities of fluid and particle
ρ_{ij}	planar spacing parameter in Figure 7.16b
σ	electrical conductivity
σ_f	free electrical surface charge
σ_Σ	surface electrical conductivity
$\tau = \varepsilon/\sigma$	charge relaxation time
$\tau_\alpha, \tau_\beta, \ldots, \tau_N$	relaxation times of layered particle
$\tau_c = \varepsilon_c/\sigma_c$	time constant used in Equation (3.12)

$\tau_m = c_m R / \sigma_c$	time constant used in Equation (3.12)
τ_{MW}	Maxwell–Wagner relaxation time defined by Equation (2.32)
τ_0	relaxation time constant defined by Equation (2.32)
τ_x, τ_y, τ_z	Maxwell–Wagner relaxation time constants along the three principal axes of ellipsoid, Equation (5.45)
ϕ	azimuthal angle in spherical coordinates
$\overline{\overline{\chi}}$	magnetic susceptibility tensor
$\chi_\parallel, \chi_\perp$	parallel and perpendicular susceptibilities of prolate spheroids, Equation (5.53)
χ^*	critical value of magnetic susceptibility in Equation (5.54a,b)
ω	radian frequency of electric or magnetic field
ω_c	critical transition frequency between + and – DEP defined by Equation (3.41)

1

Introduction

1.1 Background and motivation

Small particles in the size range from approximately one micron (10^{-6} m) up to one millimeter (10^{-3} m) are very important in today's technological world. Though often hidden from our view, they serve as tireless workhorses in many mechanisms and devices, from electrostatic copiers and printers to powder couplings to fluidized beds. Particles are used in new colloidal suspensions called *electrorheological fluids,* which respond to an applied electric field by rapidly changing their apparent viscosity. Particles are also employed in manufacturing operations including packed and fluidized bed reactors, powder coating machines, powder injection molding, etc. Many of the raw materials used in the agricultural, food, mining, and metallurgical industries are received in particulate form to be separated, beneficiated, or processed. Likewise, modern chemical technology is heavily based upon the processing of feedstocks into powdered, granular, or pelletized dry products.

Particulates, so useful and necessary in modern materials and manufacturing, can also be a nuisance or outright hazard in other situations. For example, particulate pollution is a recognized environmental and industrial health hazard. Characterization of pollutants in particulate form is an important aspect of modern environmental health science. The collection and removal of particulate matter from combustion gases is the goal of electrostatic precipitators, packed bed filters, and other pollution control apparatus. Similarly, preservation of water quality in lakes and rivers depends on removal of certain particulate matter from industrial waste water. Another example, vital in today's electronics industry, is control of submicron contaminants during fabrication and processing of solid-state devices. This hard-to-control contamination is a significant contributing factor to the high rejection rates often experienced in the fabrication of very large scale integrated (VLSI) electronic chips. Finally, airborne dust is a well-known fire and explosion hazard in certain polymer and metallurgical manufacturing operations.

One rapidly emerging branch of particulate science and technology concerns particles of biological origin, such as cells and DNA. Cells sized from less than a micron on up to several hundreds of microns make up all living organisms. The characterization, handling, and manipulation of individual cells and DNA molecules have become major thrusts of modern biomedical science and engineering. At the same time, flow cytometry has revolutionized biological assay methods by making it possible to sort and separate literally millions of cells in minutes.

Some materials technologists have labeled the decade of the 1990s the "particle age" – a fitting recognition of the tremendous advances in the manufacture of new particulate materials and the applications being discovered for them in new products and processes.

1.2 Objectives of this book

Because all particles have electrical and magnetic properties associated with their shape and with the materials of which they are constituted, they experience forces and torques when subjected to electric and/or magnetic fields. Furthermore, particles will exhibit mutual interactions – often quite strong – through the agency of their own electrical charge, polarization, or magnetization. *Particle electromechanics*, the subject of this book, may be defined as follows:

Particle electromechanics: Forces and/or torques exerted on small particles (and collections of such particles) less than approximately 10^{-3} meters in diameter through the action of an electric or magnetic field, and also the mechanics and dynamics induced by these forces and torques. The electric or magnetic field may be imposed by external means (via electrodes, magnetic pole pieces, etc.) or by other nearby charged, polarized, or magnetized particles or particle ensembles. This definition extends to the mechanics of static particle beds and to the dynamics of moving particle beds when subject to electric or magnetic fields.

The above definition encompasses the subjects of electrophoresis and dielectrophoresis; electrorheological fluids; the mechanics of electrofluidized, electrospouted, and electropacked beds; electrostatic precipitation; electrostatic particle adhesion; high-gradient magnetic separation; the magnetostabilized bed; magnetic powder couplings; magnetic field–coupled particle flow control devices; and the magnetic brush electrophotographic copier/printer. It is not the author's objective to stake out or otherwise mark such a broad ground, but rather to offer some common terminology for the subject of field–particle interactions and to provide a framework for the contributions of many engineers and researchers – past, present, and future – who work in these diverse fields.

What impresses the student of particle electromechanics is an immediately recognizable set of common phenomena manifested in diverse physical situations. For example, the same dielectrophoretic force experienced by biological cells in aqueous suspensions can also be significant in electrostatic precipitation and certain electrophotographic development processes. An analogous magnetophoretic force is exploited in high-gradient magnetic separators to filter out magnetizable particles. The exemplar of commonality is the ubiquitous phenomenon of particle chaining, which can be anticipated whenever uncharged dielectric particles, loose or in fluid suspension, are subjected to a strong electric field, or when magnetizable particles are placed in a magnetic field. Another kind of unity is found in the fundamental connection of both electromechanical forces and torques to the effective dipole moment. The point of view taken in this book is that these commonalities and interrelationships are not mere academic curiosities, but the mortar binding together a large collection of seemingly unrelated phenomena into a viable scientific and technical discipline, namely, particle electromechanics.

1.3 Limitations and caveats

As defined here, the subject of particle electromechanics can hardly be done full justice by any single volume. Therefore, certain limitations have been imposed in writing this monograph, the focus of which is the electromechanics of dielectric, conducting, and magnetizable particles in the diameter size range from about 1 μm ($\sim 10^{-6}$ m) to about 1 mm($\sim 10^{-3}$ m). This book does not supplant the late Prof. Herbert Pohl's classic text on dielectrophoresis (Pohl, 1978), but rather places that subject into a larger context. In fact, Chapter 3 covers the fundamentals of dielectrophoresis and will serve as a graduate-level introduction to the subject; however, the serious investigator of biological dielectrophoresis will be drawn inevitably to Pohl's definitive volume.

Excellent works on the physics of *electrically charged* particles are widely available in technical libraries; therefore, except in Chapter 7 where electrostatic adhesion is briefly reviewed, particle charge is not considered. The lower size limit ($\sim 10^{-6}$ m) is imposed because the mechanics of submicron particles are strongly influenced by random thermal (Brownian) motions and van der Waals forces, while the upper limit ($\sim 10^{-3}$ m) is based on a reasonable working definition of what constitutes a classical particle. We may confidently predict the rapid emergence of ultrafine particle technology and, thus, the need for a volume on the mechanics of particles smaller than 1 μm. The author of such a book will be faced with the challenging task of folding the subjects of particle electromechanics, aerosol science, and adhesion science into the unique and somewhat perplexing set of physical properties exhibited by ultrafine particles.

Another subject not covered in this book is colloidal electro-optics (and magneto-optics), which concerns the influence of electric (or magnetic) fields on the optical properties of colloidal suspensions. While particle orientation (the subject of Chapter 5) and field-induced particle chaining (discussed in Chapters 6 and 7) are electromechanical mechanisms responsible for some of the important electro-optic effects, no attention to the optics side of the problem is given here. The reader interested in electro-optics is referred to the excellent treatise on this subject by S. P. Stoylov (1991).

The subjects of electrohydrodynamics (EHD) and electroconvection, as they relate to particles, droplets, and bubbles, are not covered in the present volume. Therefore, bubble and droplet deformations induced by an electric (or magnetic) field are not considered. Likewise, no treatment of the important subject of electrophoresis of particles in aqueous suspension is provided. Only simple dielectric models for biological cells and particles in aqueous media are examined; the neglect of surface charging guarantees that the dielectric models for particles in aqueous suspension, especially biological cells, are deficient at very low frequencies. One more limitation of this book is that chain interactions among nonspherical particles have not been considered.

2

Fundamentals

2.1 Introduction

The definition of particle electromechanics offered in Chapter 1, Section 1.2, is very broad, precluding any possibility of definitive treatment in a single volume. Accordingly the scope of this book is restricted primarily to field–particle interactions involving (i) uncharged, lossy, dielectric and electrically conductive particles with AC and DC electric fields and (ii) magnetizable, electrically conductive particles with AC and DC magnetic fields. The particle electromechanics of interest here are a consequence of either the field-induced polarization of dielectric particles or the field-induced magnetization of magnetic particles. The forces and torques governing particle behavior result from the interaction of the dipole and higher-order moments with the field.

A. *Electromechanics of particles*

Two distinct types of electromechanical interactions may be identified: *imposed field* and *mutual particle* interactions. Imposed field interactions reign when a single particle, or an ensemble of noninteracting particles, is influenced by an externally imposed field. Examples include the dielectrophoretic force or the alignment torque exerted on an isolated particle. Here, it is customary to assume that the particle does not influence the field, though such an assumption is not always justified. Mutual particle interactions occur where particles are so closely spaced that the local field of a particle influences its neighbors. For particles in close mechanical contact, mutual interactions can be very strong, leading to significant changes in the equilibrium structure of particle ensembles (e.g., chain formation and cooperative electrorotation), as well as strong cohesive forces.

For the purpose of convenience in presentation, this monograph is organized into sections on imposed field interactions (Chapters 2 through 5) and mutual interactions (Chapters 6 and 7). However, the distinctions drawn between these

5

two branches of particle electromechanics are in fact artificial and have no real formal basis, because it is typical for both types of interactions to be present simultaneously. A common thread running throughout this book is the application of field–multipole interactions to calculate forces and torques on particles and particle ensembles in the cases of either imposed field or mutual interactions.

A very close analogy exists among electrostatic problems involving lossless dielectrics, DC conduction problems involving ohmic conductors, and magnetostatic problems with (linear) lossless magnetizable materials. This analogy makes it possible to start with expressions for electrical quantities such as potential and field variables, dipole moments, etc., and then to formulate the solution to the analogous conduction or magnetostatic problem by a simple change of variables. Appendix A briefly explains the origins of the analogy between electrostatic, conduction, and magnetostatic problems and contains two useful tables. Table A.1 identifies analogous quantities and their appropriate SI units, while Table A.2 summarizes the more important analogous formulas for calculation of the effective dipole moment, the gradient force, and the torque.

The approach consistently taken in this book is to cast problems in their electrostatic form first and then to consider equivalent magnetics problems second. The reader's ability to perform the appropriate changes of variables is relied upon to obtain solutions to equivalent conduction or magnetostatic problems. However, the student of particle electromechanics must also learn to recognize the limits of this analogy, which breaks down due to the properties of many common magnetizable particles. The analogy's failure should be immediately evident in the case of ferromagnetic particles exhibiting nonlinear magnetization, such as hysteresis or saturation, and permanent magnetization. But failure also occurs for the case of a conductive particle in a time-varying magnetic field, where eddy currents are induced that create a nonzero curl magnetic field within the particle. Furthermore, there is no behavior comparable to superconductivity in the realm of dielectrics.

B. Force on an infinitesimal dipole

An obvious starting point in formulating the electromechanics of particles is to estimate the net force upon a small physical dipole. This approach reveals the essential implications for particle electromechanics of the so-called ponderomotive force exerted upon dielectric materials by a nonuniform electrostatic field. The dipole consists of equal and opposite charges $+q$ and $-q$ located a vector distance $\bar{d}$ apart, and it is located in an electric field of force $\bar{E}$. Refer to Figure 2.1a. For the present, we choose to say nothing about how the electric field is created, except to stipulate that $\bar{E}$ includes no contributions due to the dipole itself.

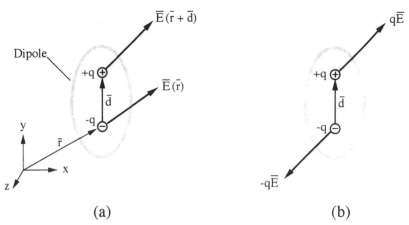

(a)   (b)

Fig. 2.1 Representation of the force and torque exerted upon a small dipole by an electric field of force.

   a) Net force on a small dipole of strength $p = qd$ in a nonuniform electric field.

   b) Coulombic force components creating net torque on a small dipole of strength $p = qd$ in a uniform electric field.

If the electric field is nonuniform, then in general the two charges ($+q$ and $-q$) will experience different values of the vector field $\bar{E}$ and the dipole will experience a net force. Performing a sum of the forces on the particle, we have

$$\bar{F} = q\bar{E}(\bar{r} + \bar{d}) - q\bar{E}(\bar{r}) \tag{2.1}$$

where $\bar{r}$ is the position vector of $-q$. Equation (2.1) can be simplified when $|\bar{d}|$ is small compared to the characteristic dimension of the electric field nonuniformity. In this case, the electric field can be expanded about position $\bar{r}$ using a vector Taylor series expansion; that is

$$\bar{E}(\bar{r} + \bar{d}) = \bar{E}(r) + \bar{d} \cdot \nabla E(\bar{r}) + \cdots \tag{2.2}$$

where all additional terms, of order d^2, d^3, and so forth, have been neglected. Using the expansion of Equation (2.2) in Equation (2.1), the following result is obtained.

$$\bar{F} = q\bar{d} \cdot \nabla \bar{E} + \cdots \tag{2.3}$$

If the limit $|\bar{d}| \to 0$ is taken in such a way that $\bar{p} \equiv q\bar{d}$ (the dipole moment) remains finite, then the following well-known expression for the force on an infinitesimal dipole results:

$$\bar{F}_{\text{dipole}} = \bar{p} \cdot \nabla \bar{E} \tag{2.4}$$

Equation (2.4) teaches that no net force is exerted on a dipole unless the externally imposed electric field is nonuniform.

The above derivation reveals that Equation (2.4) is an approximation for the force exerted upon any physical dipole, such as a polarized particle of finite size. This approximation, referred to here as the *dielectrophoretic approximation*, is usually quite adequate for imposed field interactions because the dimensions of practical electrodes are ordinarily much larger than the particles, meaning that the scale of the electric field nonuniformity is large compared to the particle dimensions. On the other hand, the DEP approximation often leads to significant error in the case of mutual interactions between closely spaced particles because the nonuniformities of the induced field due to the particles themselves are comparable to the particles' size. When this is true, Equation (2.4) becomes very inaccurate and higher-order multipolar terms must be retained. A special case where the higher-order terms can be taken into account easily is considered and analyzed in Section 2.4B below. Furthermore, in Chapters 6 and 7, which deal with mutual particle interactions, the treatment of higher-order terms is given special attention.

C. Torque on a dipole

The torque exerted by an electric field on an infinitesimal dipole may be derived by considering the net force couple acting about the center of the small dipole, as shown in Figure 2.1b. There are two contributions to this torque, one due to each charge.

$$\bar{T}^e = \frac{\bar{d}}{2} \times q\bar{E} + \frac{-\bar{d}}{2} \times (-q\bar{E}) = q\bar{d} \times \bar{E} \tag{2.5}$$

or, in terms of the dipole moment $\bar{p}$ as previously defined,

$$\bar{T}^e = \bar{p} \times \bar{E} \tag{2.6}$$

Note that the torque on the infinitesimal dipole is dependent only on the electric field vector, not its gradient, and so an electrical torque can be exerted on a dipole by a uniform field. The only requirement for the existence of this torque is that $\bar{p}$ and $\bar{E}$ are not parallel. It should be evident that, like Equation (2.4) for the dipole force, the torque expression, Equation (2.6), is an approximation for any finite dipole if the electric field is nonuniform. Any error incurred in its use becomes significant only when the scale of the electric field nonuniformities is comparable to the dimension of the dipole.

2.2 Lossless dielectric particle in an electric field

A. Induced multipolar moments

Equations (2.4) and (2.6) were derived in Section 2.1 without reference to the nature of the dipole moment $\bar{p}$. This vector quantity might be the permanent moment of a polar molecule or poled particle, or it might be induced by the imposed electric field itself. In this book, interest focuses primarily on the latter case, where the dipole moment and all higher-order moments are induced by the electric field and its derivatives. This limitation is relaxed in order to consider the mechanics of a permanently magnetized particle in a magnetic field. In general, the moment-inducing field is a combination of externally imposed and mutual field contributions (due to other nearby particles). Thus, it is necessary to relate these moments to the electric field and the particle parameters so that these values can then be used in Equations (2.4) and (2.6) to predict the forces and torques on particles.

A crucial objective of this chapter is to introduce the *effective moment* method of calculating electromechanical forces and torques exerted by electric fields upon particles. This method is based on identification of the "correct" expression for $\bar{p}$ to be used in Equations (2.4) and (2.6). Subsequent chapters on such topics as dielectrophoresis and magnetophoresis, rotation, and orientation will all employ this method whenever possible for the calculation of force and torque. The effective moment method, despite certain pitfalls to be discussed later, is valuable because it is easy to use and readily provides valid predictive results in many important cases where rigorous derivations based on the Maxwell stress tensor seem difficult or impossible.

Imagine a particle suspended in some dielectric fluid and subject to a uniform electric field. The field polarizes the particle, inducing a moment in it. The effective dipole moment $\bar{p}_{eff}$, here aligned parallel to the imposed field, is then defined as the moment of an equivalent, free-charge, point dipole that, when immersed in the same dielectric liquid and positioned at the same location as the center of the original particle, produces the same dipolar electrostatic potential. From Appendix B, the electrostatic potential Φ_{dipole} due to a point dipole of moment $\bar{p}_{eff}$ in a dielectric medium of permittivity ε_1 is

$$\Phi_{dipole} = \frac{p_{eff}\cos\theta}{4\pi\varepsilon_1 r^2} \tag{2.7}$$

where θ and r are, respectively, the polar angle and radial position in spherical coordinates. For a dielectric particle, the effective dipole moment is determined by solving the appropriate boundary value problem and then comparing the induced (dipole) term of the electrostatic potential solution to Equation (2.7). We provide an example of this procedure in the next section.

B. *Effective dipole moment of dielectric sphere in dielectric medium*

Let an insulating dielectric sphere of radius R and permittivity ε_2 be suspended in a fluid of permittivity ε_1 and be subjected to a uniform z-directed electric field of magnitude E_0. Refer to Figure 2.2. It is assumed at the outset that the electric field is uniform, that there is no free charge anywhere in the sphere or the dielectric liquid, and that the presence of the particle does not significantly disturb the system of source charges that creates E_0. The more general case of a nonuniform electric field is considered in Section 2.2E.

The electrostatic potential satisfies Laplace's equation everywhere because of the divergence- and curl-free properties of the electrostatic field. The assumed solutions for the potential outside Φ_1 and inside Φ_2 the sphere take the form

$$\Phi_1 (r, \theta) = -E_0 r \cos \theta + \frac{A \cos \theta}{r^2}, r > R \tag{2.8a}$$

$$\Phi_2 (r, \theta) = -Br \cos \theta, r < R \tag{2.8b}$$

where A and B are unknown coefficients to be determined using the boundary conditions. Note that the first term in Equation (2.8a) is the imposed uniform electrostatic field, while the second term is the induced dipole term due to the

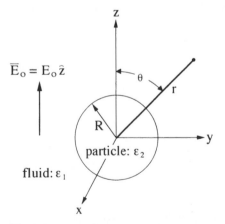

Fig. 2.2 Dielectric sphere of radius R and permittivity ε_2 immersed in a dielectric liquid of permittivity ε_1 and subjected to a uniform electric field of magnitude E_0.

particle. The boundary conditions are applied at $r = R$, the surface of the particle. First, the electrostatic potential must be continuous across the particle–fluid boundary.

$$\Phi_1(r = R, \theta) = \Phi_2(r = R, \theta) \tag{2.9a}$$

Second, the normal component of the displacement flux vector must be continuous across the boundary.

$$\varepsilon_1 E_{r1}(r = R, \theta) = \varepsilon_2 E_{r2}(r = R, \theta) \tag{2.9b}$$

where $E_{r1} = -\partial \Phi_1/\partial r$ and $E_{r2} = -\partial \Phi_2/\partial r$ are the normal electric field components in the fluid and the dielectric sphere, respectively. Combining Equations (2.8a,b) with Equations (2.9a,b), we obtain

$$A = \frac{\varepsilon_2 - \varepsilon_1}{\varepsilon_2 + 2\varepsilon_1} R^3 E_0 \quad \text{and} \quad B = \frac{3\varepsilon_1}{\varepsilon_2 + 2\varepsilon_1} E_0 \tag{2.10}$$

Comparing Equation (2.7) to the induced electric dipole term in Equation (2.8a) yields a useful general relationship between the effective moment and the coefficient A.

$$p_{\text{eff}} = 4\pi\varepsilon_1 A \tag{2.11}$$

For the special case of the homogeneous, dielectric sphere, the expression for the effective dipole moment is

$$p_{\text{eff}} = 4\pi\varepsilon_1 K R^3 E_0 \tag{2.12}$$

K, known as the Clausius–Mossotti function, provides a measure of the strength of the effective polarization of a spherical particle as a function of ε_1 and ε_2.

$$K(\varepsilon_2, \varepsilon_1) = \frac{\varepsilon_2 - \varepsilon_1}{\varepsilon_2 + 2\varepsilon_1} \tag{2.13}$$

When $\varepsilon_2 > \varepsilon_1$, then $K > 0$ and the effective moment vector $\bar{p}_{\text{eff}}$ is colinear with the imposed electric field vector $\bar{E}_0$. On the other hand, when $\varepsilon_2 < \varepsilon_1$, then $K < 0$ so that $\bar{p}_{\text{eff}}$ and $\bar{E}_0$ are antiparallel. Note that strict limits are placed upon the numerical value of K by Equation (2.13), viz. $-0.5 \leq K \leq 1.0$. Thus, even when $\varepsilon_2 \to \infty$, the magnitude of the effective moment $|\bar{p}_{\text{eff}}|$ is limited.

C. Conducting sphere in uniform DC field

The problem of a sphere of conductivity σ_2 in a fluid of conductivity σ_1 and subjected to a steady DC electric field can be worked out easily by reference to the analogy in Appendix A. The assumed solutions are the same as in Equations (2.8a) and (2.8b). The electrostatic potential remains continuous at $r = R$ so Equation (2.9a) is unchanged, while a new boundary condition replacing Equation (2.9b) expresses the continuity of the normal component of the current density at the surface of the sphere: $\sigma_1 E_{r1}(r = R, \theta) = \sigma_2 E_{r2}(r = R, \theta)$. The resulting expression for the effective moment is identical to Equation (2.12), except that now K is expressed in terms of conductivities.

$$K(\varepsilon_2, \varepsilon_1) \rightarrow K(\sigma_2, \sigma_1) = \frac{\sigma_2 - \sigma_1}{\sigma_2 + 2\sigma_1} \tag{2.14}$$

Once again, note that $-0.5 \le K \le 1.0$.

The special case of a perfectly conducting sphere, that is, $\sigma_2/\sigma_1 \rightarrow \infty$, is represented by the limit of $K = 1$, so that

$$p_{\text{eff}} \rightarrow p_0 = 4\pi\varepsilon_1 R^3 E_0 \tag{2.15}$$

Equation (2.15) is a convenient choice for normalization of the induced moments of a particle because it represents the maximum possible value of p_{eff} for a single sphere.

D. Lossless spherical shell in lossless medium with uniform field

Many commercially and industrially important types of particles start out with a layered structure or become coated during use. For example, insulating particles are sometimes treated with conductive material to reduce static electrification, while certain conductive particles are sometimes coated with insulating materials to block DC current flow. Furthermore, the various types of particles used in dry electrophotographic marking – toners and carriers – are routinely treated with compounds to improve flowability, control triboelectric charging, and achieve other desired characteristics. In other cases, particles become coated as a result of their intended use. For example, particles in granular bed filter become covered with the material they are filtering and thus exhibit changes in their dielectric properties. Biological cells are very good examples of layered particles, having an outer cell wall, insulating lipid membrane, and other complex structural features. Several simple models for cells are considered in Section 3.3E.

Such examples illustrate the need, from the standpoint of developing the subject of particle electromechanics, to be able to calculate the forces and torques exerted by electric fields upon layered particles. If the layered particle is spherically symmetric and if only the effective moment is required – e.g., for force or torque calculations – then the original particle is replaced by an equivalent homogeneous sphere of the same radius. The effective permittivity of this equivalent particle, ε_2', reflects the attributes of the original particle's structure. Consider the important case of a dielectric shell with a single concentric layer in a uniform electric field, as shown in Figure 2.3. Appendix C derives an expression for the effective permittivity ε_2' of the equivalent particle.

$$\varepsilon_2' = \varepsilon_2 \left\{ \frac{a^3 + 2\left(\dfrac{\varepsilon_3 - \varepsilon_2}{\varepsilon_3 + 2\varepsilon_2} \right)}{a^3 - \left(\dfrac{\varepsilon_3 - \varepsilon_2}{\varepsilon_3 + 2\varepsilon_2} \right)} \right\} \tag{C.4}$$

Here, R_1 is the outer radius; ε_2 is the permittivity of the layer; R_2 and ε_3 are, respectively, the radius and permittivity of the core; and $a = R_1/R_2$. Then, from Equation (C.3a), the Clausius–Mossotti function becomes

$$K(\varepsilon_2', \varepsilon_1) = \frac{\varepsilon_2' - \varepsilon_1}{\varepsilon_2' + 2\varepsilon_1} \tag{2.16}$$

The effective moment is still expressed by Equation (2.12), but now ε_2' replaces ε_2 in the expression for K. By repeated applications of Equation (C.4), starting at the innermost layer and working outward until the outer layer is reached, this substitution method may be employed to obtain an equivalent homogeneous permittivity value for a spherical particle consisting of any number of concentric dielectric layers.

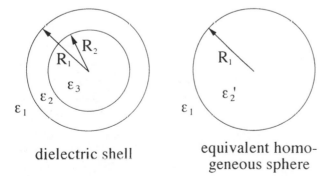

dielectric shell

equivalent homo-
geneous sphere

Fig. 2.3 Spherical concentric dielectric shell with radii R_1 and R_2 having shell permittivity ε_2 and core permittivity ε_3.

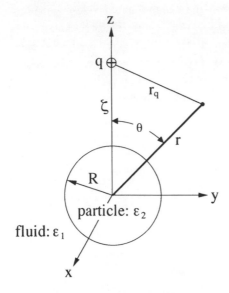

Fig. 2.4 Dielectric sphere of radius R and permittivity ε_2 immersed in a dielectric liquid of permittivity ε_1 and subjected to the electric field of a point charge q.

E. Lossless dielectric sphere in lossless dielectric medium in field of point charge

The problem of a dielectric sphere interacting with a point charge q is more general than the uniform applied field problem of Section 2.2B because the nonuniform field of the point charge induces all the linear multipolar moments – that is, the dipole, quadrupole, octupole, etc. Furthermore, the solution yields a general relationship between the induced linear moments and the various spatial derivatives of the electric field exploited in Chapters 6 and 7 to model the field interactions of closely spaced particles.

Consider the dielectric sphere shown in Figure 2.4, with radius R and dielectric permittivity ε_2, in a dielectric fluid of permittivity ε_1 and subject to the influence of a point charge q located on the z axis at distance ζ from the center of the sphere. The general solution for the axisymmetric electrostatic potential is

$$\Phi_o = \frac{q}{4\pi\varepsilon_1 r_q} + \sum_{n=0}^{\infty} \frac{A_n P_n(\cos\theta)}{r^{n+1}}, r \geq R \tag{2.17a}$$

$$\Phi_i = \sum_{n=0}^{\infty} B_n r^n P_n(\cos\theta), r < R \tag{2.17b}$$

where n is an integer and $P_n(\cos \theta)$ are the Legendre polynomials defined in Table B.1 in Appendix B. Equation (2.17a) contains the imposed field (source) term due to q and a set of induced (linear) multipolar terms. To facilitate solution for the coefficients A_n and B_n, it is convenient to expand the term $1/r_q$ in Legendre polynomials, using Equation (B.4) in Appendix B.

$$\frac{1}{r_q} = \frac{1}{\zeta} \sum_{n=0}^{\infty} \left(\frac{r}{\zeta}\right)^n P_n(\cos \theta), \ R \le r < \zeta \tag{2.18}$$

With $1/r_q$ in this form, it is now a straightforward task to invoke orthogonality plus the boundary conditions, Equations (2.9a) and (2.9b), to solve for A_n and B_n.

$$A_n = \frac{-q}{4\pi\varepsilon_1} \frac{n(\varepsilon_2 - \varepsilon_1)}{n\varepsilon_2 + (n+1)\varepsilon_1} \frac{R^{2n+1}}{\zeta^{n+1}} \tag{2.19a}$$

$$B_n = \frac{q}{4\pi} \frac{2n+1}{n\varepsilon_2 + (n+1)\varepsilon_1} \frac{1}{\zeta^{n+1}} \tag{2.19b}$$

Comparing the induced terms in Equation (2.17a) to Equation (B.6), we see that $p^{(n)} = 4\pi\varepsilon_1 A_n$, so that now we can express all the linear multipolar moment values in terms of the source charge q and other system parameters defined in Figure 2.4. However, a far more useful result may be achieved by rewriting the various moment expressions in terms of the electric field due to the charge q and its axial derivatives measured at the center of the dielectric sphere. The electric field at the particle's center due to the point charge is obtained from Coulomb's law.

$$\bar{E}(z=0) = -\left(\frac{q}{4\pi\varepsilon_1 \zeta^2}\right)\hat{z} \tag{2.20}$$

and the derivatives are

$$\frac{\partial^n \bar{E}}{\partial z^n} = -\left(\frac{(n+1)!q}{4\pi\varepsilon_1 \zeta^{n+2}}\right)\hat{z} \tag{2.21}$$

Using Equations (2.19a), (2.20), and (2.21) plus the relationship $p^{(n)} = 4\pi\varepsilon_1 A_n$ gives the following result for the linear multipolar moments in terms of the electric field and its derivatives on the central axis (Jones, 1986a).

$$p^{(n)} = \frac{4\pi\varepsilon_1 K^{(n)} R^{2n+1}}{(n-1)!} \frac{\partial^{n-1} E_z}{\partial z^{n-1}} \tag{2.22}$$

where

$$K^{(n)} = \frac{\varepsilon_2 - \varepsilon_1}{n\varepsilon_2 + (n + 1)\varepsilon_1} \tag{2.23}$$

The factors $K^{(n)}$ represent generalized polarization coefficients, of which $K^{(1)} = K$ is the Clausius–Mossotti function defined by Equation (2.13). Note that, when $n = 1$, Equation (2.22) reduces to the familiar expression for the dipole moment, that is, Equation (2.12). The higher-order moments are induced by derivatives of the electric field. For example, the linear quadrupole is induced by the first partial derivative of the field, $\partial E_z/\partial z$, while the linear octupole depends on the second derivative, $\partial^2 E_z/\partial z^2$, and so forth. Refer to Appendix B, Figure B.2, which depicts the effective point charge distributions for each of the first several linear multipoles.

Equation (2.22) has been derived for the case of a point charge but is in fact applicable for the axisymmetric electric field produced by any line charge distribution. The validity of this assertion may be proved by invoking the superposition principle and recognizing that the on-axis electric field $E_z(z)$ of any axisymmetric distribution may be synthesized by a linear array of point charges.

Stratton treats the special case of a perfectly conducting sphere in the field of a point charge separate from the dielectric sphere, but the groundwork laid by the analysis of the dielectric sphere makes such extra work unnecessary because the result for a perfectly conducting sphere can be obtained from the $\varepsilon_2/\varepsilon_1 \to \infty$ limit of an insulating dielectric sphere problem. Then, Equation (2.23) simplifies to $K^{(n)} = 1/n$ with $n = 1, 2, 3, \ldots$.

Limiting values for $K^{(n)}$ for the situation of a perfectly insulating sphere in a conductive medium (such as a bubble in a semi-insulative liquid) are obtained from the limit $\varepsilon_2/\varepsilon_1 \to 0$, in which case, $K^{(n)} = -1/(n+1)$ for $n = 1, 2, 3, \ldots$.

2.3 Dielectric particle with loss in an electric field

A variety of energy dissipation mechanisms, including conduction and dielectric relaxation phenomena, influences the behavior of real dielectric particles in an electric field. When loss is present, the dipole moment of a particle in a field exhibits either a time delay when the electric field is suddenly applied, or a phase lag when the electric field is a steady-state sinusoidal function of time. In Section 2.2, the behavior of lossless particles in an electric field was considered; in this section, the more general case of particles with ohmic loss in a sinusoidal steady-state AC electric field is examined. The transient response of a dielectric particle with ohmic loss is considered in Appendix D.

A. Homogeneous sphere in an AC electric field

We revisit the problem of an isolated homogeneous dielectric sphere (c.f., Section 2.2B) as depicted in Figure 2.2, except now assuming that the sphere and the fluid medium have finite conductivities, σ_2 and σ_1, respectively, in addition to their dielectric permittivities. The applied excitation is a uniform, linearly polarized, AC electric field of magnitude E_0 at radian frequency ω.

$$\bar{E}(t) = \text{Re}[E_0 \, \hat{z} \, \exp(j\omega t)] \tag{2.24}$$

The solution forms of Equations (2.8a,b) may be adapted by assuming that the constants in the expressions for the potential functions Φ_1 and Φ_2 are complex, that is, $A \to \underline{A}$ and $B \to \underline{B}$. The boundary condition on the potential at the particle surface, Equation (2.9a), is unchanged in its form, however, Equation (2.9b) must be replaced by a charge continuity condition because the finite conductivity permits the time-dependent accumulation of free electrical surface charge. The instantaneous charge conservation condition is

$$J_{r1} - J_{r2} + \frac{\partial \sigma_f}{\partial t} = 0, \text{ at } r - R \tag{2.25}$$

where $J_{r1} = \sigma_1 E_{r1}$ and $J_{r2} = \sigma_2 E_{r2}$ are the normal components of the ohmic current outside and inside the dielectric sphere, respectively, and σ_f is the free (unpaired) electric surface charge as defined by

$$\sigma_f = \varepsilon_1 E_{r1} - \varepsilon_2 E_{r2}, \text{ at } r = R \tag{2.26}$$

If exponential time dependence $\exp(j\omega t)$ is now assumed for all variables, we may employ the substitution $\partial/\partial t \to j\omega$. Then, the new boundary condition on the normal electric field components may be expressed in terms of complex vector phasors.

$$\underline{\varepsilon}_1 \underline{E}_{r1}(r = R, \theta) = \underline{\varepsilon}_2 \underline{E}_{r2}(r = R, \theta) \tag{2.27}$$

Equation (2.27) is of similar appearance to Equation (2.9b) but employs the familiar form of the complex dielectric constants: $\underline{\varepsilon}_1 = \varepsilon_1 + \sigma_1/j\omega$ and $\underline{\varepsilon}_2 = \varepsilon_2 + \sigma_2/j\omega$. The solution coefficients are then

$$\underline{A} = \frac{\underline{\varepsilon}_2 - \underline{\varepsilon}_1}{\underline{\varepsilon}_2 + 2\underline{\varepsilon}_1} R^3 E_0 \quad \text{and} \quad \underline{B} = \frac{3\underline{\varepsilon}_1}{\underline{\varepsilon}_2 + 2\underline{\varepsilon}_1} E_0 \tag{2.28}$$

which are identical in form to Equation (2.10), except for the replacement of scalar by complex quantities.

Using these results and the same argument employed in Section 2.2B, we now arrive at a more general expression for the complex effective moment $\bar{p}_{\text{eff}}$.

$$\bar{p}_{\text{eff}} = 4\pi\varepsilon_1 \underline{K} R^3 \bar{E}_0 \tag{2.29}$$

where the Clausius–Mossotti factor now has become a function of the complex permittivities, containing magnitude and phase information about the effective dipole moment.

$$\underline{K}(\underline{\varepsilon}_2, \underline{\varepsilon}_1) = \frac{\underline{\varepsilon}_2 - \underline{\varepsilon}_1}{\underline{\varepsilon}_2 + 2\underline{\varepsilon}_1} \tag{2.30}$$

The effective moment $\bar{p}_{\text{eff}}(t) = \text{Re}[\bar{p}_{\text{eff}} \exp(j\omega t)]$ is interpreted as the moment of a time-varying equivalent free charge dipole that produces the same electrostatic field in the same dielectric medium. Because $\underline{K}$ is complex, its magnitude $|\underline{K}|$ and phase are functions of ω. The phase angle represents the phase lag between the applied electric field and the induced moment. Such ohmic, dispersive behavior is a consequence of the finite time required to build up surface charge σ_f at the interface.

It is important to note that the factor ε_1 in Equation (2.29) for p_{eff} is *not* complex. This is so because identification of the effective moment is based upon the relationship of charge to electric field, namely, Gauss's law. It is incorrect to use the complex permittivity $\underline{\varepsilon}_1$ in the multiplier $4\pi\varepsilon_1$ appearing in Equation (2.29). The complex permittivity variables occur only in Equation (2.30) defining $\underline{K}$. As pointed out by Jones (1984) as well as Sauer (1983), several authors, making erroneous use of energy arguments, have inserted $4\pi\underline{\varepsilon}_1$ instead of the correct value $4\pi\varepsilon_1$ into dielectrophoretic force expressions. Refer to Section 2.5B for more discussion on this issue.

Equation (2.30) for complex $\underline{K}(\omega)$ is actually far more general than is suggested by its derivation here. In fact, the most general complex permittivity expressions may be substituted into Equation (2.30), that is, $\underline{\varepsilon}_1 \rightarrow \varepsilon_1' - j\varepsilon_1'' + \sigma_1/j\omega$, and $\underline{\varepsilon}_2 \rightarrow \varepsilon_2' - \varepsilon_2'' + \sigma_1/j\omega$, where ε_1'' and ε_2'' are frequency-dependent dielectric loss terms. Here, the factor $4\pi\varepsilon_1$ in Equation (2.29) is replaced by $4\pi\varepsilon_1'$.

When a lossless dielectric particle is subjected to an alternating electric field, the various moments induced in the particle can be positive (parallel to $\bar{E}_0$) or negative (antiparallel to $\bar{E}_0$), depending upon the relative magnitude of ε_1 and ε_2. The addition of dissipation adds the further complication of a phase lag between the induced moment and the applied electric field. Equation (2.30) may be rewritten to illuminate this important point.

$$\underline{K}(\omega) = \left(\frac{\sigma_2 - \sigma_1}{\sigma_2 + 2\sigma_1} \right) \left[\frac{j\omega\tau_0 + 1}{j\omega\tau_{MW} + 1} \right] \tag{2.31}$$

where

$$\tau_{MW} = \frac{\varepsilon_2 + 2\varepsilon_1}{\sigma_2 + 2\sigma_1} \quad \text{and} \quad \tau_0 = \frac{\varepsilon_2 - \varepsilon_1}{\sigma_2 - \sigma_1} \tag{2.32}$$

The characteristic relaxation time constant τ_{MW} in Equation (2.32) is the same as the relaxation time identified in the transient analysis of Appendix D. Therefore, the dispersive behavior of a homogeneous ohmic dielectric sphere in an ohmic dielectric liquid can be explained in terms of a simple, first-order, relaxation process attributable to Maxwell–Wagner interfacial polarization. Refer to Figures 2.5a and b, which provide phasor representations of the complex field and moment vectors in the complex plane for the cases of $\mathrm{Re}[\underline{K}(\omega)] > 0$ and $\mathrm{Re}[\underline{K}(\omega)] < 0$, respectively. The phase lags depicted in these phasor diagrams influence the magnitude of the dielectrophoretic force on dielectric particles (c.f., Section 3.3), and it is essential for the existence of electrical torque (c.f., Section 4.3). Equations (2.31) and (2.32) are readily shown to be consistent with earlier results of this chapter by examining the low- and high-frequency limits.

$$\lim_{\omega\tau_{MW} \to 0} [\underline{K}] = \frac{\sigma_2 - \sigma_1}{\sigma_2 + 2\sigma_1} \tag{2.33a}$$

$$\lim_{\omega\tau_{MW} \to \infty} [\underline{K}] = \frac{\varepsilon_2 - \varepsilon_1}{\varepsilon_2 + 2\varepsilon_1} \tag{2.33b}$$

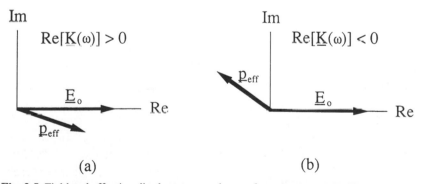

(a)   (b)

Fig. 2.5 Field and effective dipole moment phasors for two cases: a) $\mathrm{Re}[\underline{K}(\omega)]>0$ and b) $\mathrm{Re}[\underline{K}(\omega)]<0$.

The low-frequency limit, Equation (2.33a), is identical to the DC conduction result, Equation (2.14), while the high frequency limit, Equation (2.33b), is consistent with the lossless dielectric result, Equation (2.13). Furthermore, these expressions are consistent, respectively, with the $t \rightarrow \infty$ and $t \rightarrow 0$ limits of the transient problem solution given by Equation (D.5) in Appendix D.

B. Shells with ohmic loss in an AC electric field

Concentrically layered dielectric particles with ohmic loss are considered in Appendix C. If we confine attention to the effective moment of the heterogeneous particle, then the general result is obtained that a homogeneous sphere with the same radius but an effective permittivity ε_2' may be substituted for any layered sphere. The new expression for $\underline{K}$ is the expected generalization of Equation (2.16).

$$\underline{K} = \frac{\varepsilon_2' - \varepsilon_1}{\varepsilon_2' + 2\varepsilon_1} \tag{2.34}$$

For convenience, the definition of ε_2' from Appendix C is repeated below.

$$\varepsilon_2' = \varepsilon_2 \left\{ \frac{a^3 + 2\left(\dfrac{\varepsilon_3 - \varepsilon_2}{\varepsilon_3 + 2\varepsilon_2}\right)}{a^3 - \left(\dfrac{\varepsilon_3 - \varepsilon_2}{\varepsilon_3 + 2\varepsilon_2}\right)} \right\} \tag{C.6}$$

The particle, similar to that shown in Figure 2.3, has core permittivity ε_3 and core radius R_2; the concentric layer has permittivity ε_2 and outer radius R_1; and $a = R_1/R_2$. We may use Equation (C.6) repeatedly to obtain an effective permittivity value for a spherical particle with any number of layers; in general, we accrue one Maxwell–Wagner relaxation process for each interface.

From Equation (C.6), we note that the scalar values of apparent permittivity (ε_2') and conductivity (σ_2'), defined by

$$\varepsilon_2'(\omega) = \text{Re}(\varepsilon_2') \quad \text{and} \quad \sigma_2'(\omega) = -\omega \text{Im}(\varepsilon_2') \tag{C.7}$$

are dependent upon frequency and will, in general, exhibit Maxwell–Wagner polarization.

Appendix C also considers the special case of very thin surface layers, classifying them as either *series* or *shunt* admittance elements. For layers featuring series admittance, a finite potential drop may occur across the layer and the effective permittivity of the particle assumes the familiar form of the series connection of admittance terms.

$$\varepsilon_2' = \frac{\underline{c}_m R \varepsilon_2}{\underline{c}_m R + \varepsilon_2} \tag{C.10}$$

where $\underline{c}_m$ is a complex capacitance defined in conjunction with Equation (C.10) and R is the radius. In the simplest case of Equation (C.10), the surface layer is purely capacitive so that $\underline{c}_m \to c_m$, which is capacitance per unit area (F/m^2).

Another possibility is that the thin surface layer accommodates circumferential flow of real and/or displacement current around the periphery of the particle (in the $\hat{\theta}$ direction when the imposed field is directed along the z axis). In this case, a shunt admittance model is more appropriate and the effective permittivity of the particle becomes

$$\varepsilon_2' = \varepsilon_2 + \frac{2\varepsilon_\Sigma}{R} \tag{C.15}$$

Here, ε_Σ is a complex surface permittivity, defined by Equation (C.14) in Appendix C.

C. Summary

The systematic methods of Sections 2.3A and B lead to expressions for the effective moment and for the complex Clausius–Mossotti function of any layered spherical particle. Unfortunately, algebraic expressions for $\underline{K}(\omega)$ become very unwieldy for anything more complicated than a homogeneous sphere. As an alternative, a partial fraction expansion may be used for $\underline{K}(\omega)$ to simplify the mathematical representation of frequency-dependent dielectrophoretic forces and electrical torques. From Appendix E, this form is

$$\underline{K}(\omega) = K_\infty - \frac{\Delta K_\alpha}{j\omega\tau_\alpha + 1} - \frac{\Delta K_\beta}{j\omega\tau_\beta + 1} - \cdots - \frac{\Delta K_N}{j\omega\tau_N + 1} \tag{E.4}$$

where $\tau_\alpha, \tau_\beta, \cdots, \tau_N$ are the relaxation time constants of N relaxation processes inherent in the system consisting of the particle and suspension medium. In Equation (E.4), the indices $\alpha, \beta, \gamma, \ldots$ are used for the relaxation processes instead of integer indices to avoid confusion with the numbering convention chosen for the layers. Appendix E also shows that Argand diagrams of $\underline{K}(\omega)$ provide an interpretation of the frequency dependence of the dielectrophoretic force and electrical torque, which is especially useful when the relaxation processes are widely separated in the frequency spectrum, that is, when $\tau_\alpha \gg \tau_\beta \gg \cdots \gg \tau_N$. From Equation (E.4), the real and imaginary parts of $\underline{K}(\omega)$ are

$$\mathrm{Re}[\underline{K}(\omega)] = K_\infty - \frac{\Delta K_\alpha}{\omega^2 \tau_\alpha^2 + 1} - \frac{\Delta K_\beta}{\omega^2 \tau_\beta^2 + 1} - \cdots - \frac{\Delta K_N}{\omega^2 \tau_N^2 + 1} \qquad (2.35a)$$

$$\mathrm{Im}[\underline{K}(\omega)] = +\frac{\omega \tau_\alpha \Delta K_\alpha}{\omega^2 \tau_\alpha^2 + 1} + \frac{\omega \tau_\beta \Delta K_\beta}{\omega^2 \tau_\beta^2 + 1} + \cdots + \frac{\omega \tau_N \Delta K_N}{\omega^2 \tau_N^2 + 1} \qquad (2.35b)$$

Representative Re[$\underline{K}$] and Im[$\underline{K}$] for a multilayered particle depicted in Figure 2.6a with widely separated relaxations are plotted versus frequency in Figures 2.6b and c. An important distinction between these plots and the classical Debye spectra for the real and imaginary parts of dielectric constant $\underline{\kappa} = \kappa' - j\kappa''$ for a material should be noted: Re[$\underline{K}$] and Im[$\underline{K}$] can be positive or negative, while κ' and κ'' are strictly greater than zero.

In this section, by considering spherical particles with ohmic loss we have enlarged upon the concept of the effective induced moment first introduced in Section 2.2. The results derived for a homogeneous dielectric sphere with ohmic loss look similar to the expressions for an insulating particle. The only difference is that the formalism for the effective moment of a lossy dielectric sphere employs phasor variables and complex numbers. For example, compare Equa-

(a)

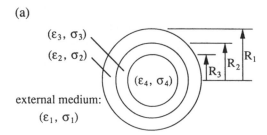

Figure 2.6 DEP and ROT spectra of a lossy dielectric particle with three layers and three distinct Maxwell–Wagner relaxations. The Argand diagram of this particle is plotted in Figure E.3a in Appendix E.

a) Multilayered dielectric shell with permittivities and radii defined as follows:
$\varepsilon_1/\varepsilon_0 = 8.0$, $\sigma_1 = 3 \bullet 10^{-6}$ S/m, $R_1 = 12\ \mu m$;
$\varepsilon_2/\varepsilon_0 = 6.0$, $\sigma_2 = 10^{-7}$ S/m, $R_2 = 10\ \mu m$;
$\varepsilon_3/\varepsilon_0 = 4.0$, $\sigma_3 = 10^{-4}$ S/m, $R_3 = 8\ \mu m$;
$\varepsilon_4/\varepsilon_0 = 2.0$, $\sigma_4 = 10^{-1}$ S/m.
where $\varepsilon_0 = 8.854 \bullet 10^{-12}$ F/m is the permittivity of free space.

b) (On facing page) Re[$\underline{K}(\omega)$] versus electrical frequency $f = \omega/2\pi$ for the multilayered particle defined in (a).

c) (On facing page) Im[$\underline{K}(\omega)$] versus electrical frequency $f = \omega/2\pi$ for the multilayered particle defined in (a).

tions (2.12) and (2.13) to Equations (2.29) and (2.30). Appendix C shows that the similarity carries over from lossless shells to shells with ohmic and/or dielectric dissipation. Therefore, by an appropriate set of variable and parameter changes

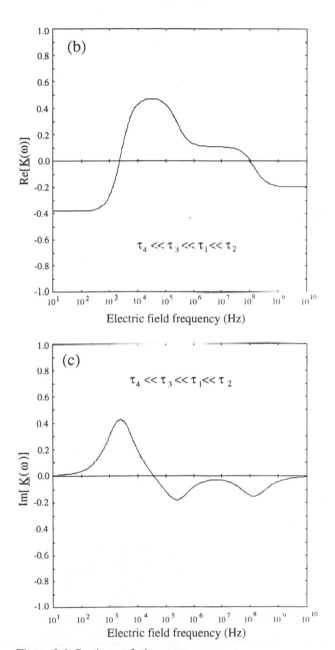

Figure 2.6 Caption on facing page.

from scalar to complex, the expressions for lossy particles can be deduced from the more readily available relations for lossless dielectrics. It should be evident from this discussion that the same set of variable changes may be incorporated into the expressions of Section 2.2E to obtain equations for the complex quadrupolar and higher-order moments for a lossy particle in a nonuniform AC electric field. Still further, the cases of nonspherical particles and chains can be dealt with using these variable changes. Equation (2.29) for the complex effective moment vector $\underline{p}_{\text{eff}}$ means that the complex Clausius–Mossotti function $\underline{K}(\omega)$ defined by Equation (2.30) or (E.4) completely determines the frequency-dependent dielectric response and the AC field electromechanics of small particles.

2.4 Effective moment calculation of force and torque

Calculation of the force and torque of electrical origin on a particle depends upon the proper identification of the effective induced moments of the particle. These moments – including the dipole and higher-order linear multipoles – are obtained by solution of a boundary value problem, as exemplified by the problem worked in Sections 2.2B and E. In this section, the assumptions inherent to this approach are examined first and then attention is devoted to the separate cases of lossless and lossy (ohmic) dielectric particles.

A. Hypothesis and definitions

The basis of the effective moment method is the hypothesis that the force and torque upon a particle can be expressed in terms of the effective moments identified from solution for the induced electrostatic field due to the particle. Consistent with this hypothesis, Equations (2.4) and (2.6) for the dipole force and torque take the following time-dependent forms.

$$\bar{F}_{\text{DEP}}(t) = \bar{p}_{\text{eff}}(t) \cdot \nabla \bar{E}(t) \tag{2.36a}$$

$$\bar{T}^{\text{e}}(t) = \bar{p}_{\text{eff}}(t) \times \bar{E}(t) \tag{2.36b}$$

Here, all variables are interpreted as instantaneous functions of time t. From simple considerations based upon Appendix B, we can extend this hypothesis to calculate the higher-order moment contributions to the force (and torque).[1] Fur-

---

[1] A rather different approach to force calculation is taken by Mognaschi and Savini (1983), who perform an integration of the Maxwell stress tensor after expanding the electric field due to a dielectric particle using Legendre polynomials.

thermore, the same effective moment methods may be invoked to calculate force and torque for nonspherical particles as well as particle chains.

The key to correct evaluation of forces and torques using the effective moment method is proper evaluation of the effective moment itself. As defined in Section 2.2A, the effective dipole $\bar{p}_{\text{eff}}$ is a free-charge, point dipole, which, when immersed in the same dielectric fluid as the original particle, produces the same dipolar field. According to this hypothesis, the instantaneous dipole moment identified by this procedure is used with Equations (2.36a,b) for calculation of instantaneous (time-dependent) force and torque. This interpretation and definition of the effective moment are not subject to criticism as long as the particle and fluid medium are lossless. However, when loss is present, conceptual difficulties arise: time or phase delays exist between the electric field $\bar{E}(t)$ and the induced dipole moment $p_{\text{eff}}(t)$, and energy arguments may not be used in support of the approach.

All higher-order moments are defined in the same way as the dipole. Thus, if $\Phi_{\text{induced},n}(r,\theta,t)$ is the nth linear multipolar component of the electrostatic potential induced by the particle, as obtained from solution of the boundary value problem, then the time-dependent effective linear multipoles $p_{\text{eff}}^{(n)}(t)$ are evaluated using

$$\Phi_{\text{induced, }n} = \frac{p_{\text{eff}}^{(n)}(t)}{4\pi\varepsilon_1 r^{n+1}} P_n(\cos\theta) \tag{2.37}$$

These moments become important in evaluation of forces between closely spaced particles.

B. Lossless particles

For a lossless particle, either homogeneous or layered, immersed in a lossless dielectric fluid of permittivity ε_1, it is a straightforward matter to identify all the linear multipolar moments $p_{\text{eff}}^{(n)}(t)$ using Equation (2.37). In the case of an AC imposed electric field, these moments will be in phase and colinear with the electric field. As for any lossless and conservative system, conventional energy arguments apply, and the electrostatic energy distribution is determined purely by the arrangement of the static electrical charges. As a result, the effect upon the energy distribution of any lossless particle is indistinguishable from that of the set of effective moments identified by the usage of Equation (2.37). Therefore, Equations (2.36a,b) unambiguously determine the dipole force and torque on the particle.

Consider first the simple case of a homogeneous dielectric sphere of radius R

and permittivity ε_2 immersed in a lossless dielectric fluid of permittivity ε_1 and subject to a slightly nonuniform electric field E_0. This electric field may be any function of time. The effective dipole moment is given by Equation (2.12) with K given by Equation (2.13), and the net force becomes

$$\bar{F}_{DEP} = 2\pi\varepsilon_1 R^3\left(\frac{\varepsilon_2 - \varepsilon_1}{\varepsilon_2 + 2\varepsilon_1}\right)\nabla E_0^2 \tag{2.38}$$

This familiar equation is the classic expression of the dielectrophoretic force exerted by a nonuniform electric field on a lossless dielectric particle (Pohl, 1951). Chapter 3 is devoted to the subject of dielectrophoresis and its phenomenology.

Equation (2.38) serves to approximate the total force on the particle if the electric field nonuniformity is modest; however, if the nonuniformity is significant, then higher-order corrections are needed. A simple way to examine the contributions of these higher-order terms to the net force is to reconsider the problem, illustrated in Figure 2.4, of a homogeneous dielectric sphere in the field of a point charge. The nonuniform field induces the set of linear multipoles $p^{(n)}$ given by Equation (2.22). This problem provides us the opportunity to check the consistency of the effective moment method of force calculation. Due to Newton's second law of motion, the vector force exerted on the point charge q by the dielectric sphere should be equal in magnitude and opposite in direction to the force exerted on the dielectric sphere by the point charge. The total force on q is unambiguously calculated using the Lorenz force law. If $\bar{E}_{induced}$ is defined as the induced field at the location of the point charge due to the sphere, then the Lorenz force acting on this charge is

$$\bar{F}_q = q\bar{E}_{induced} \tag{2.39}$$

We differentiate Equation (B.6) in Appendix B to obtain an expression for $\bar{E}_{induced}$ in terms of the effective moments induced in the sphere.

$$\bar{E}_{induced} = -\hat{z}\frac{\partial}{\partial r}\sum_{n=1}^{\infty}\frac{p_{eff}^{(n)}P_n(\cos\theta)}{4\pi\varepsilon_1 r^{n+1}}\bigg|_{r=\zeta,\theta=0} \tag{2.40}$$

Finally, we obtain

$$\bar{F}_q = \hat{z}\sum_{n=1}^{\infty}p_{eff}^{(n)}\frac{(n+1)q}{4\pi\varepsilon_1\zeta^{n+2}} \tag{2.41}$$

According to our hypothesis, the net force exerted by the charge q upon the sphere can be calculated using the effective moment method, namely, Equation (B.11) in Appendix B. The starting point is the expression for the electric field at the center of the dielectric sphere due to the point charge q, Equation (2.20), and its axial derivatives, Equation (2.21). Plugging these expressions into (B.11) gives

$$\bar{F} = \hat{z} \sum_{n=1}^{\infty} \frac{\overset{(n)}{p_{\text{eff}}}}{n!} \frac{\partial^n E_z}{\partial z^n} = -\hat{z} \sum_{n=1}^{\infty} p_{\text{eff}}^{(n)} \frac{(n+1)q}{4\pi\varepsilon_1 \zeta^{n+2}}$$

(2.42)

which is the negative of $\bar{F}_q$ from Equation (2.41).

This exercise demonstrates that the effective moment method gives the correct result for the total force at least for the simple case of a dielectric sphere in the field of a point charge. But, if we recognize that any axisymmetric electric field can be synthesized from an array of point charges, greater generality may be claimed. Using the superposition argument, we may state confidently that the validity of the effective moment force calculation method extends to all axisymmetric electrostatic field distributions. This argument, while not constituting a general proof for arbitrary electric fields, does lend considerable support to the generality claimed for the effective moment method in the lossless case.

The force exerted by an axisymmetric electric field on the nth multipolar moment may be expressed conveniently in turns of the field and its derivatives by combining Equation (2.22) with Equation (B.11) in Appendix B (Jones, 1986a).

$$F_z^{(n)} = \frac{2\pi\varepsilon_1 K^{(n)} R^{2n+1}}{n!(n-1)!} \frac{\partial}{\partial z}\left[\frac{\partial^{n-1} E_z}{\partial z^{n-1}}\right]^2$$

(2.43)

When the electric field is imposed by cylindrically symmetric electrodes, Equation (2.43) is convenient for predicting the total force. Note that the dipole ($n=1$) term is consistent with Equation (2.38) for the well-known dielectrophoretic force in an axisymmetric field.

Appendix F provides a very brief summary of a general multipolar theory for particle–field interactions. This theory, while an intuitively straightforward extension of the above formalism in the conceptual sense (Washizu and Jones, 1994), does require the use of tensor quantities to represent the multipolar moments. There are circumstances under which the general multipoles are indeed necessary; however, in most practical situations, the interaction forces favor parallel alignment of particles, in which case, the linear multipoles are adequate.

Appendix C reveals that the effective dipole moment of a dielectric shell may be determined easily based on an effective permittivity value (ε_2') for an equivalent sphere defined by Equation (C.4). However, the appendix goes on to show that the effective value of homogeneous permittivity takes on distinct different values $(\varepsilon_2)_n'$, as defined by Equation (C.19), for each of the higher-order moments. Therefore, all above expressions for the force on a homogeneous sphere may be altered for application to shells simply by making an appropriate substitution of $(\varepsilon_2)_n'$ for ε_2 within each multipolar polarization coefficient; in other words, $K^{(n)} \to K^{(n)}[(\varepsilon_2)_n', \varepsilon_1]$.

Equation (2.12) indicates that, for a lossless sphere or shell, the effective dipole moment $\bar{p}_{\text{eff}}$ is parallel to the applied electric field vector $\bar{E}_0$ (that is, $\bar{p}_{\text{eff}} \parallel \bar{E}_0$), so the dipole torque, defined by Equation (2.36b), is identically zero. In general, the effective moment will be nonparallel to the electric field only if (i) the particle has ohmic or dielectric loss leading to a phase lag between the field and the induced moment or (ii) the particle exhibits material or shape-dependent anisotropy. Refer respectively to Chapters 4 and 5, which treat these two situations.

C. Particles with ohmic (or dielectric) loss

If the particle or the medium in which it is suspended exhibits loss, then, strictly speaking, potential energy arguments are no longer valid and the effective moment method of force calculation is left without a rigorous theoretical foundation. Nevertheless, we propose here to use the effective multipole method because of its ease of use, because of its capability of getting answers when more rigorous methods fail, and because of the demonstrated validity of its answers in the cases that can be tested. In Section 2.5, we compare force and torque equations obtained using the effective moment method to a rigorously derived result based on the Maxwell stress tensor (Sauer, 1985; Sauer and Schlögl, 1985).

The hypothesis exemplified in Equations (2.36a,b) is that the correct effective moment to be used in the equations for force and torque is the instantaneous value identified from solution of the relevant boundary value problem. Consider once again a homogeneous sphere with permittivity ε_2 and conductivity σ_2 in a fluid of permittivity ε_1 and conductivity σ_1 and subject to a slightly nonuniform, sinusoidally time-varying electric field. According to the above interpretation of the effective moment method, the time-dependent expression for the force and torque will be

$$\bar{F}_{\text{DEP}}(t) = \text{Re}[\bar{p}_{\text{eff}}\exp(j\omega t)] \cdot \nabla\text{Re}[\bar{\bar{E}}\exp(j\omega t)] \tag{2.44a}$$

$$\bar{T}^e(t) = \text{Re}[\bar{p}_{\text{eff}}\exp(j\omega t)] \times \text{Re}[\bar{\bar{E}}\exp(j\omega t)] \tag{2.44b}$$

These equations indicate that the force and torque on a particle in an AC electric field will consist of two components: a constant (average) term and a time-varying term at twice the electric field frequency. For particles in the size range from 1 to 1000 μm, any motion due to the time-varying term is usually damped heavily by the viscosity of the suspension medium, so that only the time-average term is important. Nevertheless, double-frequency components are present and observable when using low-frequency electric fields in DEP experiments.

From Equation (2.44a), the time-average dielectrophoretic force may be written in terms of the complex expressions for the dipole moment and the electric field.

$$\langle \bar{F}_{DEP}(t) \rangle = \frac{1}{2} \mathrm{Re}[\bar{p}_{eff} \cdot \nabla \underline{E}^*] \tag{2.45}$$

where $\langle f(t) \rangle$ represents the time-average of the function $f(t)$ and the asterisk signifies complex conjugation. Using Equation (2.29) to replace the complex effective dipole moment, we have

$$\langle \bar{F}_{DEP}(t) \rangle = 2 \pi \varepsilon_1 R^3 \mathrm{Re}[\underline{K}(\omega)] \nabla E_{rms}^2 \tag{2.46}$$

where E_{rms}, still a function of position, is the root-mean-square magnitude of the imposed AC electric field. Therefore, the real part of the complex Clausius–Mossotti function determines the frequency dependence of the average dielectrophoretic force. This frequency dependence, which depends upon the parameters of the particle and the suspending medium, defines the DEP spectrum. Refer to the discussion of dielectrophoretic levitation in Section 3.4 in Chapter 3. Certain generic attributes of the frequency dependence of $\mathrm{Re}[\underline{K}(\omega)]$ are examined in Appendix E.

We may obtain a more general multipolar force expression by straightforward extension of the above derivation to Equation (2.43).

$$\langle F_z^{(n)} \rangle = \frac{2 \pi \varepsilon_1 R^{2n+1}}{n!(n-1)!} \mathrm{Re}[\underline{K}^{(n)}] \frac{\partial}{\partial z} \left[\frac{\partial^{n-1} E_{z,rms}}{\partial z^{n-1}} \right]^2 \tag{2.47}$$

As anticipated, the frequency dependence of the higher-order force components is determined by the real part of the complex polarization coefficient $\underline{K}^{(n)}$.

Using similar arguments, Equation (2.44b) may be used to obtain an expression for the time-average electric torque in terms of the complex expressions for the dipole moment and the electric field phasor.

$$\langle \bar{T}^e(t) \rangle = \frac{1}{2} \mathrm{Re}[\bar{p}_{eff} \times \underline{E}^*] \tag{2.48}$$

The conventional approach to calculating the electrical torque is to assume that the electric field is a right circularly polarized vector.

$$\bar{E} = E_0 (\hat{x} - j\hat{y}) \tag{2.49}$$

where E_0 is the magnitude. Combining Equations (2.29) and (2.49) with (2.48), the result is

$$\langle \bar{T}^e(t) \rangle = -4\pi\varepsilon_1 R^3 \mathrm{Im}[\underline{K}(\omega)] E_0^2 \hat{z} \tag{2.50}$$

Thus, the frequency dependence of the electrical torque is determined by the imaginary part of the complex polarization coefficient $\underline{K}$, just as the real part of $\underline{K}$ determines the dielectrophoretic force. The frequency-dependent behavior of $\mathrm{Im}[\underline{K}(\omega)]$ is discussed at length in Chapter 4 and Appendix E.

Appendix C reveals that all effective multipolar moments of a lossless dielectric shell may be determined easily using the effective permittivity values $(\varepsilon_2)_n'$ in Equation (C.19) for the polarization coefficient $K^{(n)}$. It will be evident that the methods employed to obtain Equation (C.19) may be generalized to account for ohmic loss through appropriate replacement of scalar permittivities by their complex counterparts. In the case of lossy shells, we may use the substitution: $(\varepsilon_2)_n' \rightarrow (\underline{\varepsilon}_2)_n'$ where

$$(\underline{\varepsilon}_2)_n' = \underline{\varepsilon}_2 \left\{ \frac{a^{2n+1} + (n+1)\underline{K}^{(n)}}{a^{2n+1} - n\underline{K}^{(n)}} \right\} \tag{2.51}$$

Thus, the above force and torque expressions for a lossy, homogeneous sphere may be adapted for lossy shells by the substitutions $\underline{\varepsilon}_2 \rightarrow (\underline{\varepsilon}_2)_n'$ in $\underline{K}^{(n)}$, or wherever the complex particle permittivity appears. Extension of the force equations (2.46) and the more general (2.47) to shells and other layered spherical particles should be obvious by this procedure. The same substitutions may be used with Equation (2.50) to calculate the electrical torque on a lossy shell.

2.5 Theory of Sauer and Schlögl

In 1985, Sauer and Schlögl published the first rigorous derivation of the dielectrophoretic force and electrical torque on lossy spheres (Sauer, 1985; Sauer and Schlögl, 1985). Their analysis was based upon a Maxwell stress tensor formulation specially adapted for AC sinusoidal steady-state conditions within lossy media (Meixner, 1961). The stress tensor was integrated over the spherical surface of the particle, necessitating a cumbersome expansion of the slightly non-uniform electric field which significantly complicated the integration. The mathematical difficulties were such that attention was restricted to the simple case of a homogeneous sphere immersed in a dielectric liquid and subject to a slightly nonuniform AC electric field. Both ohmic conduction and *dielectric loss* were included in the constitutive law models for the particle and liquid.

A. *Summary of their result*

The principal results of Sauer and Schlögl are formulas for the time-average force and torque exerted by electric fields upon homogeneous spheres. When written in the nomenclature of this book, their expressions are very similar to Equations (2.46) and (2.50), except that $\underline{\varepsilon}_1$ and $\underline{\varepsilon}_2$ are replaced by more general complex permittivities to account for intrinsic dielectric loss mechanisms.

$$\langle \bar{F}_{\mathrm{DEP}}(t) \rangle = 2\pi\varepsilon_1' R^3 \mathrm{Re}[\underline{K}(\omega)] \, \nabla E_{\mathrm{rms}}^2 \qquad (2.52a)$$

$$\langle \bar{T}^e(t) \rangle = -4\pi\varepsilon_1' R^3 \mathrm{Im}[\underline{K}(\omega)] E_0^2 \qquad (2.52b)$$

where the substitutions

$$\underline{\varepsilon}_1 \rightarrow \varepsilon_1' \; -\mathrm{j}\varepsilon_1'' + \frac{\sigma_1}{\mathrm{j}\omega} \quad \text{and} \quad \underline{\varepsilon}_2 \rightarrow \varepsilon_2' \; -\mathrm{j}\varepsilon_2'' + \frac{\sigma_2}{\mathrm{j}\omega} \qquad (2.53)$$

have been made in the complex Clausius–Mossotti function $\underline{K}$. The only new elements are the components $\mathrm{j}c_1''$ and $\mathrm{j}c_2''$, representing dielectric loss mechanisms inherent in the particle and the liquid suspension medium. Note that, in general, all the permittivity terms (ε_1', ε_1'', ε_2', and ε_2'') will depend upon frequency. When $\varepsilon_1'' = \varepsilon_2'' = 0$, Sauer and Schlögl's force and torque equations are in complete agreement with results obtained using the effective moment method for homogeneous spheres with ohmic loss.

B. *Reconciliation with effective moment method*

The most important question that arises is how to adapt the effective moment method to account correctly for the case of dielectric loss. From the standpoint of the boundary value solution methodology used to identify an effective moment $\underline{p}_{\mathrm{eff}}$, there are no conceptual or theoretical difficulties. The substitutions of (2.53) may be performed wherever the complex permittivities $\underline{\varepsilon}_1$ and $\underline{\varepsilon}_2$ appear in the boundary value solutions summarized in Section 2.3A. Then, the effective dipole moment

$$\underline{p}_{\mathrm{eff}} = 4\pi(\underline{\varepsilon}_1' \; - \mathrm{j}\underline{\varepsilon}_2')\underline{K}R^3 E_0 \qquad (2.54)$$

correctly predicts the measurable induced potential external to the particle. The problem arises in reconciling this expression for the effective moment to the force and torque expressions of Sauer and Schlögl. Working backward from their expressions, one infers a slightly different expression for the complex effective moment of a lossy homogeneous sphere.

$$(\underline{p}_{eff})_{force} = 4\pi\varepsilon_1'\underline{K}R^3E_0 \tag{2.55}$$

In both Equations (2.54) and (2.55), $\underline{K}$ is determined using the substitutions of Equation (2.53); that is

$$\underline{K} \rightarrow \frac{(\varepsilon_2' - j\varepsilon_2'' - j\sigma_2/\omega) - (\varepsilon_1' - j\varepsilon_1'' - j\sigma_1/\omega)}{(\varepsilon_2' - j\varepsilon_2'' - j\sigma_2/\omega) + 2(\varepsilon_1' - j\varepsilon_1'' - j\sigma_1/\omega)} \tag{2.56}$$

Therefore, to obtain the correct force and torque expressions, we must ignore one component (the term proportional to $j\varepsilon_1''$) of the effective moment phasor in Equation (2.54).

In comparing Equation (2.54) to Equation (2.55), one notes that dielectric loss ε_2'' due to the particle presents no such theoretical difficulties in effective moment identification, because it influences the force and torque values only through $\underline{K}^{(n)}$. This is not surprising because, to an observer outside the particle, a volume relaxation process intrinsic to the material of the particle is indistinguishable from some Maxwell–Wagner surface polarization mechanism attributable to purely ohmic loss.

A number of previous works have incorrectly invoked the energy method to obtain the force expression for a sphere (and the alignment torque expression for an ellipsoid) immersed in liquids with ohmic loss. In all these cases, the error is indiscriminate use of the substitutions of (2.53) in the force or torque equations derived for the lossless case. The expressions resulting from such substitutions are not consistent with the equations of Sauer and Schlögl; furthermore, even when dielectric loss is absent, they do not reduce to the correct answer in the DC (low-frequency) limit (Jones, 1984). Therefore, we see that it is the presence of dielectric loss in the suspension medium that leads to ambiguity in proper identification of $\underline{p}_{eff}$.

C. Discussion

The effective moment method, as defined in Section 2.2A, may be used confidently to calculate forces and torques as long as dielectric loss in the suspension liquid in negligible, that is, if $\varepsilon_1'' \ll \varepsilon_1'$. If dielectric loss cannot be ignored, the effective moment method still is salvageable if a new set of relationships replaces Equation (2.11) for identification of the complex effective moment $\underline{p}_{eff}$.

$$\underline{p}_{eff} = 4\pi (\varepsilon_1' - j\varepsilon_1'') \underline{A} \tag{2.57}$$

and

$$(\underline{p}_{eff})_{force} = 4\pi\varepsilon_1' \underline{A} \tag{2.58}$$

$\underline{p}_{eff}$, as defined in Equation (2.57), is used for determining the strength of induced electric fields, while $(\underline{p}_{eff})_{force}$, defined by Equation (2.58), is used in Equations (2.45) and (2.48) for force and torque calculations.

From the standpoint of classical electrodynamics, the rigorous derivations of DEP force and torque performed by Sauer and Schlögl (1985) are very important contributions. They provide the solid theoretical underpinning for dielectrophoretic phenomena so long awaited. At the same time, the derivations themselves are conceptually difficult to understand, largely due to the use of nonphysical (that is, unmeasurable) pressure-like variables. Furthermore, the need to perform difficult integrals over the surface of the particle in effect restricts solutions to simple spherical shapes situated in weakly nonuniform electric fields. Therefore, this theory has the disadvantage of being very hard to apply to some of the most important practical cases of particle electromechanics, such as nonspherical particles, interactions of closely spaced particles, particle chains, etc. On the other hand, the effective moment method, despite its nonrigorous origins, is easy to use and thus fills a serious need for a method to obtain expressions for forces and torques when use of the Maxwell stress tensor is not possible.

In this book, the principal approach to the calculation of forces and torques on particles is the effective moment method, but with due care taken to avoid the difficulties associated with dielectric loss in the medium. In many circumstances, such loss is negligible; therefore few difficulties with the effective moment method are anticipated. For the important case of biological cells suspended in aqueous media containing nutrients, dielectric loss can become important, especially at higher electrical frequencies (greater than ~100 MHz). In these situations, the use of $(\underline{p}_{eff})_{force}$ as defined by Equation (2.58) should give correct predictions for DEP forces and torques.[2]

---

[2] Clearly, more theoretical work is needed to extend the Sauer–Schlögl theory and to provide more insight into the relationship of effective moments to induced fields and measurable electromechanical effects.

3

Dielectrophoresis and magnetophoresis

3.1 Introduction

In the last chapter, we used a dipole model to provide a physical interpretation for the forces and torques exerted upon small particles by an electric field. An effective moment methodology was advanced for calculation of the dipole and higher-order multipolar forces. We extended this method to heterogeneous particles, that is, concentrically layered spherical shells, and to particles with dielectric and/or ohmic losses. With this necessary modeling framework now at our disposal, we direct our attention to the phenomenology of dielectrophoresis. *Dielectrophoresis (DEP)* refers to the force exerted on the induced dipole moment of an uncharged dielectric and/or conductive particle by a nonuniform electric field. This book's early focus on DEP serves as acknowledgment that the dipole force term predominates over higher-order multipolar components in the electromechanics of particles except in the case of a particle located near a field null or in the strongly nonuniform electric field of another closely spaced particle. Dielectrophoresis is technologically important in its own right, as evidenced by the number of applications in such scientific and technical fields as biophysics, bioengineering, electrorheological fluids, and mineral separation.[1]

Electromechanical forces are exerted upon magnetic particles in a nonuniform magnetic field as well. This phenomenon, called *magnetophoresis (MAP)*, is to some extent analogous to dielectrophoresis; however, because of the nonlinearity exhibited by most magnetic materials and the effectively diamagnetic behavior of (i) conductive particles in AC magnetic fields and (ii) superconductive particles in DC magnetic fields, there are some important differences. For these reasons, we treat the subject of magnetophoresis separately in Section 3.5 and highlight certain limitations of the dielectric–magnetic analogy.

---

[1] For a thorough treatment of biological dielectrophoresis, the reader is referred to the definitive volume on this subject, written by H. A. Pohl (1979).

A. Phenomenological definition

The term "dielectrophoresis" was coined by H. A. Pohl (1951), who performed important early experiments with small plastic particles suspended in insulating dielectric liquids and found that the particles would move in response to the application of a nonuniform AC or DC electric field. Recognizing that new terminology was needed to describe the phenomena under study, he combined the word for force, "phoresis" from the Greek, with the word "dielectric" to arrive at this new term. He intended the new word to describe the force exerted on uncharged dielectric particles by virtue of their polarizability. The phenomenological bases of his definition are catalogued below:

- Particles experience a DEP force only when the electric field is nonuniform.
- The DEP force does not depend on the polarity of the electric field and is observed with AC as well as DC excitation.
- Particles are attracted to regions of stronger electric field when their permittivity ε_2 exceeds that of the suspension medium ε_1, i.e., when $\varepsilon_2 > \varepsilon_1$.
- Particles are repelled from regions of stronger electric field when $\varepsilon_2 < \varepsilon_1$.
- DEP is most readily observed for particles with diameters ranging from approximately 1 to 1000 μm.

In Section 3.2, we show that the expressions for the DEP force derived in Chapter 2 are consistent with Pohl's observations. Then, in Section 3.3, we extend the predictive dielectrophoretic theory to the case of particles with frequency-dependent effective dielectric polarizability in AC electric fields.

B. Further delineation of dielectrophoretic effect

Throughout this book, the term "dielectrophoresis" is used consistently to encompass two fundamentally related but observably distinct types of forces. These are: (i) the forces exerted upon individual noninteracting particles by an externally imposed nonuniform electric field and (ii) the mutual attractive or repulsive force between two or more closely spaced particles due to the moments induced by an external electric field. In the latter case, the externally imposed electric field may be uniform or nonuniform. In general, field interactions involving higher-order multipolar moments are included within this definition; however, from a practical standpoint, the DEP term often dominates. Inclusion of mutual interactions in an expanded definition of dielectrophoresis is natural and amounts to a very modest extension of Pohl's original intent.[2]

---

[2] Occasionally, the term "dielectrophoresis" is employed in reference to the collection and confinement of insulating dielectric liquids using nonuniform electric fields (Melcher et al., 1969). [Footnote continued on p. 36]

The value of a word diminishes when its definition becomes too inclusive. Therefore, the author proposes that several manifestations of particle electromechanics – specifically particle rotation, orientation, and electrophoresis – be clearly distinguished from the definition of dielectrophoresis. Particle rotation and orientation, the subjects of Chapters 4 and 5, respectively, can be thought of as imposed field–dipole moment interactions, but the resulting torque leads to rotational or orientational effects, rather than the particle translation phenomena studied by Pohl. Furthermore, particle rotation and/or orientation are important enough to merit their own distinct terminology, despite the fact that they are often observed to occur together with dielectrophoresis.

Electrophoresis is a strong, polarity-dependent phenomenon related to double-layer formation at the surface of certain particles suspended in aqueous media. The effect manifests itself as an apparent particle charge, does not require electric field nonuniformity, and should not be confused with DEP. The charge is of electrochemical origin; ions of one sign collect at the particle's surface while a surrounding counter-ion cloud shields this net charge from the bulk of the fluid. When a DC electric field is applied, the mobile counter-ion charge is set in motion, creating a fluid flow parallel to the surface. The particle moves in response to the shear stresses exerted upon the particle by this fluid convection. The AC electrophoretic effect is confined to very low frequencies. Readers interested in electrophoresis should consult the excellent book by V. G. Levich (1962) for in-depth coverage of the motion of particles in electrolytic solutions.

3.2 DEP phenomenology for lossless spherical particle

A. Dielectrophoretic force expression

From Equation (2.38), the net DEP force acting upon a lossless dielectric sphere of permittivity ε_2 and radius R, suspended in a medium of permittivity ε_1 and subjected to an electric field E_0, is

$$\bar{F}_{\mathrm{DEP}} = 2\pi\varepsilon_1 R^3 K \nabla E_0^2 \qquad (3.1)$$

where, for convenience, the expression for the force has been rewritten using the Clausius–Mossotti function K.

$$K = \frac{\varepsilon_2 - \varepsilon_1}{\varepsilon_2 + \varepsilon_1} \qquad (3.2)$$

Such usage, while not consistent with Pohl's definition, has achieved common acceptance and is not particularly misleading. On the other hand, the indiscriminate use of this term to describe interfacial electrohydrodynamic (EHD) phenomena exhibited by dielectric liquids is inappropriate and best avoided.

Equation (3.1) is quite general and may be used for any spatially dependent electric field vector $\bar{E}_0(\bar{r})$ such that $|\bar{E}_0| = E_0$.

B. Phenomenology of DEP

A close examination of Equation (3.1) reveals all the important features of the dielectrophoretic effect exhibited by lossless dielectric particles in lossless media. For reference, these are enumerated below:

(i) F_{DEP} is proportional to particle volume.

(ii) F_{DEP} is also proportional to ε_1, the dielectric permittivity of the medium in which the particle is suspended.

(iii) The DEP force vector is directed along the gradient of the electric field intensity ∇E_0^2, which, in general, is not parallel to the electric field vector $\bar{E}_0(\bar{r})$.

(iv) The DEP force depends upon the magnitude and sign of K, the Clausius–Mossotti function.

The above list demonstrates that Equation (3.1) is in complete agreement with Pohl's phenomenological definition of dielectrophoresis as summarized in Section 3.1Λ.

Consistent with (iii) and (iv) and with Pohl, we choose to distinguish between *positive* and *negative* dielectrophoretic effects.

Positive dielectrophoresis: $K > 0$ (or $\varepsilon_2 > \varepsilon_1$). Particles are attracted to electric field intensity maxima and repelled from minima.

Negative dielectrophoresis: $K < 0$ (or $\varepsilon_2 < \varepsilon_1$). Particles are attracted to electric field intensity minima and repelled from maxima.

Figures 3.1a and b illustrate the positive and negative dielectrophoretic effects for two distinct electrode geometries. In the concentric cylindrical geometry of Figure 3.1a, the observed force is parallel to the vector electric field, that is, $\bar{F}_{DEP}$ ‖ $\bar{E}_0$. On the other hand, between the nonparallel electrode plates shown in Figure 3.1b, the DEP force and the electric field vector are orthogonal, that is, $\bar{F}_{DEP}$ ⊥ $\bar{E}_0$. In the general case, the force vector $\bar{F}_{DEP}$ can have any orientation with respect to the electric field vector $\bar{E}_0$.

For a layered dielectric particle or shell, the essential phenomenology is unchanged. It is only necessary to replace ε_2 by ε_2' in K, where ε_2' is the effective permittivity value of the layered particle as defined by Equation (C.4) in Appendix C. With this substitution ($\varepsilon_2 \rightarrow \varepsilon_2'$), the definitions of positive and negative dielectrophoresis are unaltered.

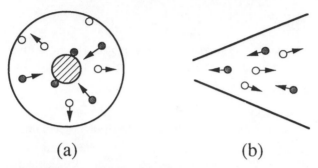

● particles exhibiting +DEP effect
○ particles exhibiting -DEP effect

(a)   (b)

Fig. 3.1 Positive and negative dielectrophoretic effects in different electrode structures.
(a) Electric field parallel to its gradient: $\bar{F}_{DEP} \parallel \bar{E}_0$
(b) Electric field vector perpendicular to its gradient: $\bar{F}_{DEP} \perp \bar{E}_0$

One rather significant feature of the DEP force is that it is strongly dependent upon the geometry of the electrodes used to create the nonuniform electric field. This sensitivity may be expressed through the following scaling law (Bahaj and Bailey, 1979).

$$|\bar{F}_{DEP}| \propto V^2/L_e^3 \tag{3.3}$$

where V is the applied voltage and L_e is a characteristic length for the electrodes. This scaling law reveals that the magnitude of the dielectrophoretic force on a particle can be increased by reducing the electrode dimensions. Furthermore, both the voltage and the electric field magnitudes required to achieve a fixed value of the DEP force for a particle diminish as the size of geometrically similar electrodes is decreased. This feature of the scaling law may be exploited to minimize joule heating when using dielectrophoretic forces to manipulate cells or other particles suspended in aqueous media.

C. *Higher-order multipolar force terms*

While Equation (3.1) includes only the dipole force, the general definition of dielectrophoresis adopted in Section 3.1B actually encompasses general multipolar contributions of all orders. These terms become very important in the case of closely spaced interacting particles, where dipole interactions are simply overwhelmed by higher-order effects. Recall that Section 2.4 contains an analysis based upon a linear multipolar expansion, useful for the special case of an axi-symmetric field. Interacting particles aligned with their line of centers parallel to the applied field meet this condition and are considered at length in Chapters 6 and 7. Refer to Appendix F, which briefly summarizes a general multipolar theory for dielectrophoresis that accounts for all multipoles.

Examination of Equation (F.5) for $\bar{F}_{total}$, the general expression for total force on a spherical dielectric particle, reveals that the sign of all force terms still depends upon the relative values of ε_1 and ε_2. In particular, the nth multipolar term is attracted toward maxima or minima of the $(n-1)^{th}$ derivative of the field, depending on whether $\varepsilon_2 > \varepsilon_1$ or $\varepsilon_2 < \varepsilon_1$, respectively. Furthermore, the dependence of the various terms of the series upon radius R signifies that simple proportionality to particle volume is not preserved for the higher-order force contributions. All this apparent complexity should not obscure the fundamental physical principle that a small dielectric particle always seeks a location to maximize or minimize the square of the electric field magnitude integrated over the volume of the particle, depending on whether $\varepsilon_2 > \varepsilon_1$ or $\varepsilon_2 < \varepsilon_1$, respectively.

3.3 Frequency-dependent DEP phenomenology

The fundamental phenomenology of dielectrophoresis, as codified by Pohl (1951), is correctly predicted for lossless dielectric particles suspended in lossless media by Equation (3.1). However, the most interesting dielectrophoretic behavior is exhibited by particles exhibiting frequency-dependent dispersion due to dielectric or conductive loss. Such frequency-dependent dielectrophoretic phenomena are the subject of this section.

A. Homogeneous sphere with ohmic loss

Consider the problem of a homogeneous dielectric particle with ohmic conductivity but no dielectric loss suspended in a dielectric fluid medium, also with purely ohmic loss. Since dielectric loss is ignored (that is, $\varepsilon_1'' = \varepsilon_2'' = 0$), then Sauer and Schlögl's basic expression for the time-average DEP force, Equation (2.52a), becomes

$$\langle \bar{F}_{DEP} \rangle = 2\pi \varepsilon_1 R^3 \text{Re}[\underline{K}(\omega)] \nabla E_{rms}^2 \tag{3.4}$$

and, starting from Equation (2.56), the complex Clausius–Mossotti function $\underline{K}(\omega)$ is

$$\underline{K} = \frac{\varepsilon_2 - \varepsilon_1 - j(\sigma_2 - \sigma_1)/\omega}{\varepsilon_2 + 2\varepsilon_1 - j(\sigma_2 + 2\sigma_1)/\omega} \tag{3.5}$$

All frequency dependence of the time-average DEP force $\langle \bar{F}_{\mathrm{DEP}} \rangle$ is contained within the factor Re[$\underline{K}$]. A very convenient form nicely illuminating this frequency dependence is (Benguigui and Lin, 1982)

$$\mathrm{Re}\,[\underline{K}] = \frac{\varepsilon_2 - \varepsilon_1}{\varepsilon_2 + 2\varepsilon_1} + \frac{3\,(\varepsilon_1\sigma_2 - \varepsilon_2\sigma_1)}{\tau_{\mathrm{MW}}(\sigma_2 + 2\sigma_1)^2(1 + \omega^2\tau_{\mathrm{MW}}^2)} \tag{3.6}$$

where τ_{MW} is the same Maxwell–Wagner charge relaxation time used in Chapter 2, Equation (2.31), as well as Appendix D, Equation (D.3).

$$\tau_{\mathrm{MW}} = \frac{\varepsilon_2 + \varepsilon_1}{\sigma_2 + 2\sigma_1} \tag{3.7}$$

This time constant characterizes the decay of a dipolar distribution of free charge at the surface of the sphere.

The familiar high- and low-frequency limits for Re[$\underline{K}$] may be identified in terms of $\omega\tau_{\mathrm{MW}}$.

$$\mathrm{Re}\,[\underline{K}] \;\rightarrow\; \begin{cases} \dfrac{\sigma_2 - \sigma_1}{\sigma_2 + 2\sigma_1}, \text{ for } \omega\tau_{\mathrm{MW}} \ll 1 \\[2mm] \dfrac{\varepsilon_2 - \varepsilon_1}{\varepsilon_2 + 2\varepsilon_1}, \text{ for } \omega\tau_{\mathrm{MW}} \gg 1 \end{cases} \tag{3.8}$$

DC conduction governs the low-frequency DEP behavior, and dielectric polarization governs the high-frequency behavior. Figures 3.2a, and b plot Re[$\underline{K}$] versus normalized radian frequency $\omega\tau_{\mathrm{MW}}$ for two different cases. In case a, where $\varepsilon_2/\varepsilon_1 > 1$ and $\sigma_2/\sigma_1 < 1$, Re[$\underline{K}$] is negative at low frequencies and positive at high frequencies. Just the reverse is assumed in case b, where $\varepsilon_2/\varepsilon_1 < 1$ and $\sigma_2/\sigma_1 > 1$, so that Re[$\underline{K}$] becomes positive at low frequencies and negative at high frequencies.

The frequency dependence of Re[$\underline{K}$] translates to a frequency-dependent, time average DEP force on a homogeneous particle. In other words, both the magnitude and sign of $\langle \bar{F}_{\mathrm{DEP}} \rangle$ are functions of the electric field frequency ω. For example, if $\varepsilon_2 > \varepsilon_1$ and $\sigma_2 < \sigma_1$, the particle will be attracted to electric field intensity minima at low frequencies and to maxima at high frequencies. This frequency-dependent behavior reverses if $\varepsilon_2 < \varepsilon_1$ and $\sigma_2 > \sigma_1$. Frequency-selective action at a distance has been exploited in mineralogical beneficiation (Benguigui et al., 1986; Shalom and Lin, 1988), biological assays performed on living cells (Pohl and Crane, 1971; Kaler and Pohl, 1983; Price et al., 1988), as well as the dielectrophoretic levitator (Tombs and Jones, 1991). DEP levitation, with

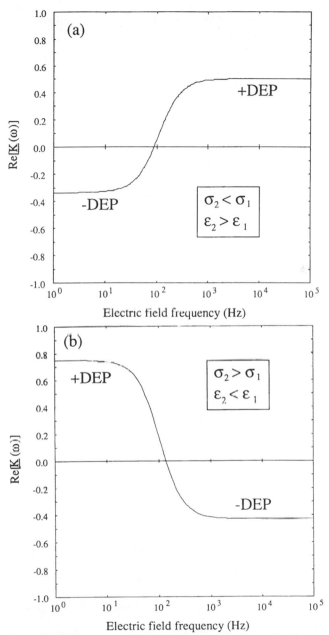

Fig. 3.2 Dielectrophoretic spectra of homogeneous dielectric spheres with ohmic loss.

(a) Re[$\underline{K}$] plot for $\varepsilon_1/\varepsilon_0 = 2.5$, $\varepsilon_2/\varepsilon_0 = 10.0$, $\sigma_1 = 4 \cdot 10^{-8}$ S/m, $\sigma_2 = 10^{-8}$ S/m, and $R = 5$ μm versus frequency in Hz.

(b) Re[$\underline{K}$] plot for $\varepsilon_1/\varepsilon_0 = 10.0$, $\varepsilon_2/\varepsilon_0 = 1.0$, $\sigma_1 = 10^{-8}$ S/m, $\sigma_2 = 10^{-7}$ S/m, and $R = 5$ μm versus frequency in Hz.

applications in the characterization and remote manipulation of small particles in the 5- to 500-μm-diameter size range, is the subject of Section 3.4.

B. Time-dependent DEP

The time-dependent effective moment $p_{eff}(t)$ of a homogeneous ohmic dielectric sphere subjected to a suddenly applied DC electric field of magnitude E_0 is derived in Appendix D. We may employ this result to obtain an expression for the time-dependent dielectrophoretic force upon the particle. Combining Equations (2.36a) and (D.3) yields

$$\bar{F}_{DEP}(t) = 2\pi\varepsilon_1 R^3 \left(\frac{\sigma_2 - \sigma_1}{\sigma_2 + 2\sigma_1} \right) [1 - \exp(-t/\tau_{MW})] \nabla E_0^2 \tag{3.9}$$

$$+ 2\pi\varepsilon_1 R^3 \left(\frac{\varepsilon_2 - \varepsilon_1}{\varepsilon_2 + 2\varepsilon_1} \right) \exp(-t/\tau_{MW}) \nabla E_0^2, t \geq 0$$

Equation (3.9) describes the time dependence of the DEP force, which can change magnitude and sign depending upon the relative magnitudes of the permittivities and conductivities. This capability of achieving both positive and negative dielectrophoretic effects on the same particle in a given fluid suspension medium on different time scales has been investigated for possible application in new mineral separation and beneficiation processes (Benguigui and Lin, 1984). Note that when $\varepsilon_1/\sigma_1 = \varepsilon_2/\sigma_2$, all time dependence of the force vanishes, just as does all frequency dependence of Re[$\underline{K}(\omega)$].

C. Layered spherical shells with loss

The dispersive behavior of layered spherical particles with loss is complicated by the interfacial (Maxwell–Wagner) polarization occurring at each layer boundary. Nonetheless, the systematic methods of Appendix C may be employed to model such particles when loss is present. Equations (C.6) and (C.8) are used with Equation (2.44a) for the time-average force to predict the frequency-dependent dielectrophoretic response.

For any layered particle, the time-average DEP force is given by (3.4), using

$$\underline{K} = \frac{\varepsilon_2' - \varepsilon_1'}{\varepsilon_2' + \varepsilon_1'} \tag{3.10}$$

along with Equation (C.6) for ε_2'. Equation (3.10) incorporates the influences of each layer and, with one relaxation process identified at each boundary, the frequency-dependent dispersion of a particle can be very complex. Fortunately, the simple form for $\underline{K}(\omega)$ provided by Equation (E.4) in Appendix E may be exploited to reveal all the relaxation frequencies. By combining Equations (2.35a) and (2.46) in Chapter 2, we obtain a compact expression for the frequency-dependent, time-average DEP force on a layered spherical particle.

$$\langle \bar{F}_{DEP}(t) \rangle = 2\pi\varepsilon_1 R^3 \left[K_\infty - \frac{\Delta K_\alpha}{\omega^2 \tau_\alpha^2 + 1} - \frac{\Delta K_\beta}{\omega^2 \tau_\beta^2 + 1} - \cdots - \frac{\Delta K_N}{\omega^2 \tau_N^2 + 1} \right] \nabla E_{rms}^2 \qquad (3.11)$$

D. Generalized definition for +DEP and –DEP

For layered, lossy particles, the magnitude and sign of $\langle \bar{F}_{DEP} \rangle$ are controlled by the real part of $\underline{K}(\omega)$, as now defined by Equation (3.10) or its convenient alternative, Equation (E.4). We may thus generalize the definitions for positive and negative dielectrophoresis of Section 3.2B for any particle, layered or homogeneous, lossless or dispersive.

Positive dielectrophoresis: $\text{Re}[\underline{K}(\omega)] > 0$. Particles are attracted to electric field intensity maxima and repelled from minima.

Negative dielectrophoresis: $\text{Re}[\underline{K}(\omega)] < 0$. Particles are attracted to electric field intensity minima and repelled from maxima.

Appendix C provides all the necessary means to evaluate the effective permittivity ε_2' and therefore $\underline{K}(\omega)$ for any concentrically layered spherical particle. Though the complex algebra involved often presents formidable obstacles to obtaining simple analytical expressions, symbolic manipulation software and computer languages supporting complex variables and operations are now readily available to perform necessary computations.

Though not proven here, the above definitions for positive and negative DEP in terms of $\text{Re}[\underline{K}(\omega)]$ remain valid for nonspherical particles if suitable generalization of $\underline{K}(\omega)$ is made to account for shape factors. These shape factors may be derived from solution of Laplace's equation in an approach analogous to that taken in Section 2.2B. The important special case of lossy ellipsoidal particles is considered in Chapter 5.

E. Examples of lossy spherical shells

The greater complexity of the frequency-dependent DEP response of lossy shells is best illustrated by specific examples chosen from some of the simpler dielectric models for biological cells, namely, walled cells and protoplasts.

E.1 Model of walled cell

Walled cells are the structural building blocks of plants, but many important single-cell microorganisms such as yeasts also take a basically similar form. The very simple spherical model shown in Figure 3.3a depicts three distinct regions of a walled cell: the wall itself, the cell membrane, and the cell interior. The wall is an exostructure that helps provide rigidity to the cell. For example, plant tissue consists of large numbers of these cells held together by the wall material, which gives structural rigidity to leaves, stems, and other plant parts. Yeasts are walled, single-cell organisms that usually assume a roughly spherical shape. In yeasts, the cell wall provides mechanical protection to the fragile membrane. The cell wall is usually modeled as a homogeneous spherically concentric shell of finite thickness with bulk permittivity ε_w and ohmic conductivity σ_w.

The membrane, which consists of a selectively permeable bi-layer of lipid protein molecules, is thin ($\sim 10^{-2}\,\mu m$) and quite fragile. It serves a vital function as a two-way conduit for (i) life-sustaining nutrients and regulatory substances required by the cell's metabolism and (ii) waste materials excreted from the cell. Membrane structure and function are subjects of intense investigation in bio-medical science, but attention here is restricted to the membrane's average dielectric properties, which are remarkable in themselves. Typically, membranes can withstand a DC electric potential drop up to ~ 1 V without sustaining damage. This breakdown voltage corresponds to an effective dielectric strength of $\sim 10^8$ V/m, a value sustainable in few if any synthetic dielectric insulation materials. The membrane behaves like a very low loss capacitor, blocking low-frequency electric fields and electric current from the interior of the cell. In accordance with the surface admittance concept introduced in Appendix C, membranes are typically characterized by effective capacitance c_m and conductance g_m, both per unit surface area.

The cell interior is extraordinarily complex, containing the nucleus and vacuoles as well as numerous other structures, all suspended in an aqueous ionic fluid called the cytoplasm. Despite this complexity, simplified models for the interior can be used in many circumstances. In particular, we use here a homogeneous model for the cytoplasm with dielectric permittivity ε_c and ohmic conductivity σ_c.

Using the methodology of Appendix C, the walled cell is replaced by an equivalent homogeneous sphere with complex permittivity $\underline{\varepsilon}_2' = \varepsilon_2' + \sigma_2'/j\omega$, where $\underline{\varepsilon}_2'$ and σ_2' are both frequency-dependent. The complex effective permittivity $\underline{\varepsilon}_2'$ is obtained by sequential use of Equation (C.10) and then (C.6). Because it is usually impossible to obtain any sort of decipherable analytical result for either $\underline{\varepsilon}_2'$ or $\underline{K}(\omega)$ from these operations, the best strategy is to employ

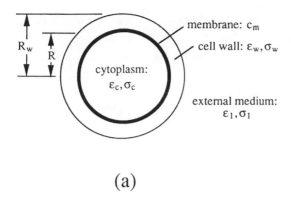

(a)

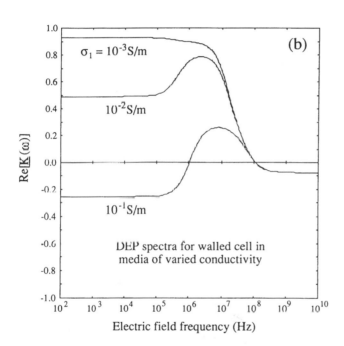

Fig. 3.3 Simple dielectric model of walled (yeast) cell. Refer to Figure E.5 (Appendix E) for Argand plot of $\underline{K}(\omega)$ for this model cell using identical parameters.

(a) Definition of walled cell modeling parameters.

(b) Re[$\underline{K}$] versus frequency in Hz for the following model parameters:

cytoplasm: $\varepsilon_c/\varepsilon_0 = 60.0$, $\sigma_c = 0.5$ S/m, $R = 2.0$ μm;

lossless membrane: $c_m = 1.0$ μF/cm^2, $g_m = 0$;

cell wall: $\varepsilon_w/\varepsilon_0 = 65.0$, $\sigma_w = 0.1$ S/m, $R_w = 2.5$ μm;

suspension medium: $\varepsilon_1/\varepsilon_0 = 78.0$, $\sigma_1 = 10^{-3}, 10^{-2}, 10^{-1}$ S/m.

computer methods to study the frequency dependence. Figure 3.3b shows a plot of Re[$\underline{K}$] versus frequency for a set of parameters typical of common brewers' and bakers' yeast (*S. cerevisiae*) in aqueous suspension media of electrical conductivity σ_1 covering the range routinely used in DEP experiments. In Figure 3.3b, several relaxation frequencies are evident and, furthermore, the DEP response is strongly dependent upon σ_1. For the higher values of σ_1, negative DEP reigns at low frequencies. This is a consequence of the DC current blocking effect of the highly insulating membrane. As σ_1 is lowered, the high effective particle conductivity of the particle σ_2' dominates in the low-frequency range and positive DEP takes over. At high frequencies the membrane becomes electrically transparent, revealing the interior of the cell to possess dielectric permittivity very similar to that of the suspension medium. As a consequence, $|\text{Re}[\underline{K}(\omega \to \infty)]| \ll 1$ and the DEP effect becomes weak.

E.2 Protoplast model

Protoplasts are very fragile, balloon-like particles prepared by treating walled cells with special enzymes to digest the wall. Refer to Figure 3.4a, which shows the simple dielectric model for a protoplast: a conductive fluid interior enclosed by a very thin capacitive layer. Mammalian cells, most of which lack a cell wall, are structurally somewhat comparable and, though smaller and less fragile, exhibit a polarization response similar to protoplasts.

The dielectric model of a protoplast is simpler than the walled cell, and analytical expressions can be used to represent both the effective particle permittivity ε_2' and $\underline{K}(\omega)$ in certain limits. The expression for the effective permittivity is

$$\varepsilon_2' = c_m R \left[\frac{j\omega\tau_c + 1}{j\omega(\tau_m + \tau_c) + 1} \right] \tag{3.12}$$

The quantities c_m and R have already been defined for Figure 3.4a, while $\tau_m = c_m R/\sigma_c$ and $\tau_c = \varepsilon_c/\sigma_c$. Furthermore, we have assumed here that the transmembrane conductance g_m is negligible. Using Equation (3.12), the complex Clausius–Mossotti function $\underline{K}(\omega)$ becomes

$$\underline{K}(\omega) = -\frac{\omega^2(\tau_1\tau_m - \tau_c\tau_m') + j\omega(\tau_m' - \tau_1 - \tau_m) - 1}{\omega^2(\tau_c\tau_m' + 2\tau_1\tau_m) - j\omega(\tau_m' + 2\tau_1 + \tau_m) - 2} \tag{3.13}$$

Here, $\tau_1 = \varepsilon_1/\sigma_1$ and $\tau_m' = c_m R/\sigma_1$. To determine the frequency dependence of the DEP force acting on a protoplast, it is necessary to obtain the real part of Equation (3.13). Figure 3.4b plots Re[$\underline{K}$] for a set of parameters typical of a plant protoplast, once again with suspension conductivity varied to show its influence on the spectra.

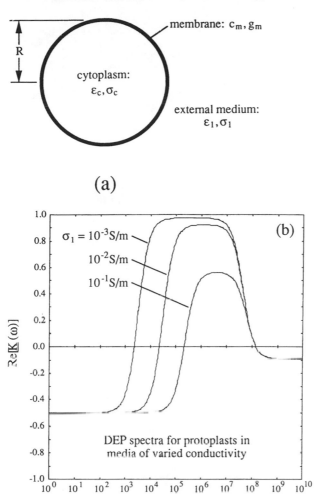

(a)

Fig. 3.4 Dielectric model of protoplast. Refer to Figure E.4a for the Argand plot of $\underline{K}(\omega)$ for this model cell using the same parameters.

(a) Definition of protoplast model parameters.

(b) Re[$\underline{K}$] versus frequency in Hz for the following model parameters:

cytoplasm: $\varepsilon_c/\varepsilon_0 = 60.0$, $\sigma_c = 0.5$ S/m, $R = 2.0$ μm;

lossless membrane: $c_m = 1.0$ μF/cm^2, $g_m = 0$;

suspension medium: $\varepsilon_1/\varepsilon_0 = 78.0$, $\sigma_1 = 10^{-3}$, 10^{-2}, 10^{-1} S/m.

Note three regions of the frequency spectrum where Re[$\underline{K}$] is roughly constant, separated by two distinct breakpoints. This behavior has a straightforward physical interpretation revealed by a close examination of Equation (3.13). The low- and high-frequency limits, so clearly evident in Figure 3.4b, are

$$\underline{K}(\omega) = \begin{cases} -0.5 \text{ for } \omega \to 0 \\ \dfrac{\varepsilon_c - \varepsilon_1}{\varepsilon_c + 2\varepsilon_1} \text{ for } \omega \to \infty \end{cases} \tag{3.14}$$

The low-frequency limit is due to the membrane, which blocks DC current and causes the cell to behave like an insulating sphere. At high frequencies, the membrane is electrically transparent, making the particle behave like a homogeneous dielectric sphere of permittivity ε_c.

Figure 3.4b depicts a flat portion of the Re[$\underline{K}$] curve at intermediate frequencies that is fairly broad when the suspension conductivity is low. An approximate expression for Re[$\underline{K}(\omega)$] can be obtained from Equation (3.13) for this regime.

$$\text{Re}[\underline{K}(\omega)] \approx \frac{\tau_m' - (\tau_1 + \tau_m)}{\tau_m' + 2(\tau_1 + \tau_m)} \tag{3.15}$$

It is usual for protoplasts that $\tau_1 \gg \tau_m$, so that Equation (3.15) may be further simplified.

$$\text{Re}[\underline{K}(\omega)] \approx \frac{c_m R - \varepsilon_1}{c_m R + 2\varepsilon_1} \tag{3.16}$$

Equation (3.16) lends itself to an appealing physical interpretation: at intermediate frequencies, the cytoplasm behaves like a perfect conductor, effectively shorting the cell interior and concentrating the entire potential drop across the cell at the membrane. As a result, the cell's polarization is completely dominated by the membrane capacitance c_m. From Equation (3.16), the effective permittivity of the cell ε_2' is approximately equal to $c_m R$. Typically, $c_m R \sim 5000\varepsilon_0$ and $\varepsilon_1 = 78\varepsilon_0$, so that Re[$\underline{K}$] ≈ 1.0.

Plots of the type shown in Figures 3.3b and 3.4b provide important information about the DEP spectra, that is, the magnitude and sign of the frequency-dependent dielectrophoretic force exerted by a nonuniform AC electric field upon heterogeneous particles. But other graphic forms may be used to present this frequency dependence; one such form, the Argand diagram described in Appendix E, uses the complex plane to visualize the trajectory of $\underline{K}(\omega)$ as the frequency is swept from low to high values. Argand plots provide a composite representation of the DEP and rotational spectra as well as the influence of important particle and suspension parameters upon them. Argand plots obtained with the same parameters used to obtain Figures 3.3b and 3.4b are provided, respectively, in Figures E.5 and E.4 in Appendix E.

3.4 DEP levitation

Dielectrophoretic levitation fulfills a somewhat specialized need among the scientific and technical applications for dielectrophoresis. The theory of DEP levitation developed below builds upon the definitions of *positive* and *negative* dielectrophoresis propounded in Section 3.2B (and generalized in Section 3.3D). We proceed by first deriving a useful theorem for electrostatic fields and then engaging it to establish the requirements for suspending a dielectric particle in fluid with the dielectrophoretic force. Two types of levitation – *passive* and *feedback-controlled* – may be used to suspend particles exhibiting, respectively, negative and positive DEP behavior.

A. DEP levitation theory

Consider the electrostatic field $\bar{E}(x,y,z)$ externally imposed upon a homogeneous, isotropic, linear dielectric continuum containing no volume-distributed unpaired electric charge. Such a vector field must have zero divergence.

$$\nabla \bullet \bar{E} = 0, \quad \text{or} \quad \frac{\partial E_i}{\partial x_i} = 0 \tag{3.17}$$

Furthermore, this field must have zero curl.

$$\nabla \times \bar{E} = 0, \quad \text{or} \quad \frac{\partial E_i}{\partial x_j} = \frac{\partial E_j}{\partial x_i} \tag{3.18}$$

in order to satisfy the conventional electrostatic approximation. Equations (3.17) and (3.18) express the divergence and curl conditions in vector notation on the left and indicial notation on the right. Here, i and j are indices referring to the three coordinate axes, that is, $x = x_1$, $y = x_2$, and $z = x_3$. Furthermore, the Einstein summation convention is used in Equation (3.17). According to this convention, $E^2 = E_i E_i$ and

$$\nabla^2 E^2 = \frac{\partial}{\partial x_j}\frac{\partial}{\partial x_j} E_i E_i \tag{3.19a}$$

which may be reduced to

$$2\frac{\partial E_i}{\partial x_j}\frac{\partial E_i}{\partial x_j} + 2E_i\frac{\partial}{\partial x_j}\left(\frac{\partial E_i}{\partial x_j}\right) \tag{3.19b}$$

The second term in Equation (3.19b) is exactly zero. To show this result, use the curl-free condition, Equation (3.18)

$$E_i \frac{\partial}{\partial x_j} \left(\frac{\partial E_i}{\partial x_j} \right) = E_i \frac{\partial}{\partial x_j} \left(\frac{\partial E_j}{\partial x_i} \right) \tag{3.20}$$

Then, change the order of the partial derivatives and use the divergence condition, Equation (3.17).

$$E_i \frac{\partial}{\partial x_j} \left(\frac{\partial E_j}{\partial x_i} \right) = E_i \frac{\partial}{\partial x_i} \left(\frac{\partial E_j}{\partial x_j} \right) = 0 \tag{3.21}$$

From Equations (3.19a,b) and (3.21) we finally obtain an inequality for the Laplacian of the field intensity squared.

$$\nabla^2 E^2 \geq 0 \tag{3.22}$$

This inequality establishes the important result for DEP levitation that maxima of the electric field intensity, isolated from electrodes or any discontinuity within the suspension medium, cannot exist within divergence- and curl-free electrostatic fields (Jones and Bliss, 1977). On the other hand, examples of electrostatic fields with isolated minima (including zeros) abound. Consider for example the quadrupolar field, which has a field intensity zero at the origin.

To levitate a particle in a conservative force field $\bar{F}$ at some point (x_0, y_0, z_0) called P, an equilibrium condition is required:

$$\bar{F}(x_0, y_0, z_0) = 0 \tag{3.23}$$

where $\bar{F}$ is the net vector force experienced by the particle. If we assume that this net force is conservative, then

$$\bar{F} = -\nabla \Psi \tag{3.24}$$

where Ψ is a scalar function representing the net electromechanical potential. Combining Equations (3.23) and (3.24), we have

$$\nabla \Psi \big|_{x_0, y_0, z_0} = 0 \tag{3.25}$$

Stability of the equilibrium defined by Equation (3.23) is guaranteed if the force on the particle is nonzero and inward-directed at all points on the closed surface Σ defining an arbitrarily small volume δV enclosing the point P. This condition may be expressed in terms of the normal derivative of the potential function Ψ evaluated on the surface.

$$\left.\frac{\partial \Psi}{\partial n}\right|_{\Sigma} > 0 \tag{3.26}$$

Refer to Figure 3.5 for the definitions of $\hat{n}$ the outward-directed unit vector, and Σ, the closed surface enclosing the point P. From Green's theorem (Hildebrand, 1962), we know that

$$\oint_{\Sigma} \frac{\partial \Psi}{\partial n} ds = \int_{\delta V} \nabla^2 \Psi dV \tag{3.27}$$

Therefore, by combining Equations (3.26) and (3.27), we may impose the following requirement for stability upon the Laplacian of Ψ at the point (x_0, y_0, z_0) (Lin and Jones, 1984):

$$\left.\nabla^2 \Psi\right|_{x_0, y_0, z_0} > 0 \tag{3.28}$$

For dielectrophoretic levitation of a lossless dielectric particle in a gravitational field, the net force is

$$\bar{F}_{net} = -2\pi R^3 \varepsilon_1 \frac{\varepsilon_2 - \varepsilon_1}{\varepsilon_2 + 2\varepsilon_1} \nabla F_0^2 + m_b \bar{g} \tag{3.29}$$

Here, m_b is the buoyant mass, which may be positive or negative, and $\bar{g}$ is the gravitational acceleration vector, which is assumed to be spatially uniform. From the definition of the mechanical potential Ψ, Equation (3.24), we have

$$\nabla^2 \Psi = -2\pi R^3 \varepsilon_1 \frac{\varepsilon_2 - \varepsilon_1}{\varepsilon_2 + 2\varepsilon_1} \nabla^2 E_0^2 \tag{3.30}$$

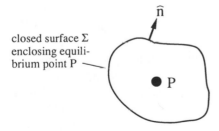

closed surface Σ
enclosing equili-
brium point P —

Fig. 3.5 Equilibrium point P with closed surface Σ defining small volume δV and unit vector $\hat{n}$.

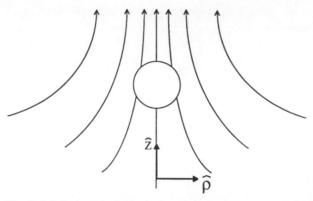

Fig. 3.6 Spherical particle of radius R on the axis of cusped, axisymmetric electrostatic field with coordinates z and ρ defined. Electrodes are not shown.

The only way to satisfy Equation (3.28), the condition for stable confinement of a particle, and Equation (3.22), the condition required of all divergence-free electrostatic fields, is for $\varepsilon_2 < \varepsilon_1$, or $K < 0$. In the case of a lossy particle or shell, this condition is generalized to $\mathrm{Re}[\underline{K}(\omega)] < 0$. If we then define *passive dielectrophoretic levitation* as the stable confinement of a dielectric particle by a nonuniform electrostatic field (without resort to position sensing and feedback), then the following theorem may be stated:

Stable passive dielectrophoretic levitation of a particle is possible only if the particle exhibits a negative dielectrophoresis effect, that is, $\varepsilon_2 < \varepsilon_1$ for a lossless homogeneous particle, $\varepsilon_2' < \varepsilon_1$ for a lossless shell, or $\mathrm{Re}[\underline{K}(\omega)] < 0$ for the more general case of a particle or shell with loss.

In Section 3.4D, we show that levitation of particles exhibiting positive dielectrophoresis is also possible but requires position sensing and a feedback loop, plus an electric field with specific spatial attributes.

B. Dynamic model of DEP levitator

We build the dynamic levitator model upon the simplifying assumption that the electrodes and the field they produce are axisymmetric.[3] Refer to Figure 3.6, which illustrates in section view a cusped, cylindrically symmetric, electrostatic field with a spherical particle positioned on the axis. Some practical electrode designs for passive and feedback-controlled DEP levitators are described later in this section. Generalized linear equations of motion for the particle near the equilibrium point are derived below in order to explore the governing stability

---

[3] Experiments with a new class of passive DEP levitators featuring azimuthally periodic, planar electrodes have been reported (Washizu and Nakada, 1991; Fuhr et al., 1992). A generalized analysis, which assumes azimuthal periodicity, has been reported recently (Jones and Washizu, 1994).

conditions. This approach allows us to identify the different conditions required for both passive and feedback-controlled levitation.

Consider an electrostatic field produced by a pair of cylindrically symmetric electrodes immersed in a fluid of dielectric constant ε_1 and to which the voltage $V(t)$ has been applied. The reasonable assumption is made that any particle equilibria, if they exist at all, will be located on the z axis. The curl and divergence conditions, Equations (3.17) and (3.18), may be used to obtain two-dimensional Taylor series forms for the axial (E_z) and radial (E_ρ) components of the field near the axis (Jones, 1981).

$$E_z = V\,[\alpha_0 + \alpha_1 z + \alpha_2(z^2 - \rho^2/2) + \cdots] \tag{3.31a}$$

$$E_\rho = V\,[-\alpha_1 \rho/2 - \alpha_2 \rho z + \cdots] \tag{3.31b}$$

where ρ is the radial coordinate of the particle. Note that $E_\rho = 0$ on the axis, that is, at $\rho = 0$. The coefficients α_0, α_1, α_2, etc., depend upon the geometry and spacing of the electrodes. They are functions of the axial position and are related to the various axial derivatives of the axial field E_z.

$$\alpha_n(z=z_0) \equiv \frac{1}{Vn!} \left.\frac{\partial^n E_z}{\partial z^n}\right|_{z=z_0,\,\rho=0} \tag{3.32}$$

Equation (3.32) can be used with either known analytical solutions or numerical values for the field of any cylindrically symmetric electrode geometry to calculate the α_n coefficients. These coefficients control the stability of all particle equilibria.

The equations governing axial and radial motion for a levitated particle must account for the gravitational and dielectrophoretic forces as well as any viscous damping that may be present. From Newton's first law,

$$m_{\text{eff}}\ddot{z} = -m_b g - b\dot{z} + 2\pi R^3 \varepsilon_1 \text{Re}[\underline{K}(\omega)]\,\frac{\partial E^2}{\partial z} \tag{3.33a}$$

$$m_{\text{eff}}\ddot{\rho} = -b\dot{\rho} + 2\pi R^3 \varepsilon_1 \text{Re}[\underline{K}(\omega)]\,\frac{\partial E^2}{\partial \rho} \tag{3.33b}$$

where m_{eff} is the effective mass of the particle in the fluid, m_b is the buoyant mass, $g = 9.81$ m/s^2 is the acceleration due to gravity in the $-z$ direction, and b, always positive, is the coefficient of velocity-dependent viscous damping.[4] In

---

[4] For a spherical particle, $m_{\text{eff}} = 4\pi R^3(\rho_2 + \rho_1/2)/3$ and $m_b = 4\pi R^3(\rho_2 - \rho_1)/3$, where ρ_1 and ρ_2 are the mass densities of the fluid and the particle, respectively (Lamb, 1945, Sec. 92). The Stokes drag formula for a sphere moving through a viscous liquid gives $b = 6\pi\eta_1 R$, where η_1 is the dynamic viscosity.

writing these equations, we have used the time-average expression for the DEP force on a lossy dielectric sphere in an AC electric field, that is, Equation (3.4). For this approach to be valid, the viscous damping or the particle's inertia must be sufficiently strong so that the particle does not respond perceptibly to the double-frequency, zero time-average component of the DEP force. Even in relatively nonviscous liquids, the damping is usually quite strong for particles smaller than ~20 µm in diameter at the frequencies above ~10 Hz. With larger particles or very low frequencies, this motion can be a problem unless the viscosity is much higher. For example, the minimum usable AC electric field frequency for ~50-µm-diameter glass beads suspended in 10-centistoke silicone oil is approximately 20 Hz (Tombs and Jones, 1991).

The next step is to linearize Equations (3.33a,b) by making the following set of definitions:

$$z(t) \equiv z'(t), \ \left| z'/L_e \right| \ll 1 \tag{3.34a}$$

$$\rho(t) \equiv \rho'(t), \ \left| \rho'/L_e \right| \ll 1 \tag{3.34b}$$

$$V(t) \equiv V_0 + v'(t), \ \left| v'/V_0 \right| \ll 1 \tag{3.34c}$$

where L_e is the characteristic dimension of the electrodes. Because the choice of the origin is arbitrary, we suffer no loss of generality assuming that $z_0 = 0$. The primed variables z' and ρ' signify small displacements of the particle from an equilibrium assumed to exist on the axis at the point $z = 0$, $\rho = 0$ for $V = V_0$. Next, the field expansions of Equations (3.31a,b) are employed to reduce the equations of motion (3.33a,b) to an equilibrium condition

$$0 = -m_b g + 4\pi R^3 \varepsilon_1 \text{Re}[\underline{K}(\omega)] \alpha_0 \alpha_1 V_0^2 \tag{3.35}$$

and a set of coupled second-order linear perturbation equations in z', ρ', and v'

$$\ddot{z}' + \frac{b}{m_{\text{eff}}} \dot{z}' - \frac{4\pi R^3 \varepsilon_1 \text{Re}[\underline{K}] V_0^2}{m_{\text{eff}}} \left[\alpha_1^2 + 2\alpha_0 \alpha_2 \right] z' - \frac{8\pi R^3 \varepsilon_1 \text{Re}[\underline{K}] V_0}{m_{\text{eff}}} \left[\alpha_0 \alpha_1 \right] z' = 0$$

$$\tag{3.36a}$$

$$\ddot{\rho}' + \frac{b}{m_{\text{eff}}} \dot{\rho}' - \frac{2\pi R^3 \varepsilon_1 \text{Re}[\underline{K}] V_0^2}{m_{\text{eff}}} \left[\frac{\alpha_1^2}{2} - 2\alpha_0 \alpha_2 \right] \rho' = 0 \tag{3.36b}$$

It must be kept in mind that α_0 and α_1 are functions of the chosen equilibrium particle position z_0.

These perturbation equations may be used to investigate stability of the equilibria defined by Equation (3.35). First, we direct attention to the case of passive levitation, where the electric field distribution maintains the particle in stable equilibrium with no outside intervention and where the applied voltage is fixed; that is, $v' = 0$. For successful levitation, the particle must be stable with respect to axial and radial excursions:

for axial stability,

$$\left[\alpha_1^2 + 2\alpha_0\alpha_2 \right] \text{Re}[\underline{K}] < 0 \tag{3.37a}$$

for radial stability,

$$\left[\alpha_1^2 - 4\alpha_0\alpha_2 \right] \text{Re}[\underline{K}] < 0 \tag{3.37b}$$

C. Passive DEP levitation

We examine the cases of positive and negative dielectrophoresis separately. When $\text{Re}[\underline{K}] > 0$, the required conditions on the axisymmetric electrostatic field are, from Equations (3.37a,b),

$$\alpha_1^2 < -2\alpha_0\alpha_2 \quad \text{and} \quad \alpha_1^2 < 4\alpha_0\alpha_2 \tag{3.38}$$

These two inequalities cannot be satisfied simultaneously and therefore passive dielectrophoretic levitation of particles exhibiting positive DEP cannot be achieved in an axisymmetric field. This result is just as expected from the theorem derived in Section 3.4A.

When $\text{Re}[\underline{K}] < 0$, the conditions imposed upon the axisymmetric electrostatic field coefficients are

$$\alpha_1^2 > -2\alpha_0\alpha_2 \quad \text{and} \quad \alpha_1^2 > 4\alpha_0\alpha_2 \tag{3.39}$$

which can be satisfied simultaneously. Therefore, passive dielectrophoretic levitation in an axisymmetric field of particles exhibiting negative DEP may be achieved, once again as anticipated from Section 3.4A.

C.3 Electrode configurations

The behavior of a particle in a passive DEP levitator depends on the nature of the electrostatic field imposed by the electrodes. For all axisymmetric structures, the locus of levitation points is a finite or semi-infinite line segment on the axis. This locus is delimited by stronger and weaker electric field end points that are defined by marginal conditions from one of three distinct types. The stronger field limit, called the *release point*, is the location on the axis where the axial derivative of E_z^2 reaches its maximum value, that is, $\alpha_1^2 = -2\alpha_0\alpha_2$. When a particle passes the release point, it cannot be recaptured by the DEP force and is lost. The weaker field limit can be either a *field null* or a point of *marginal radial stability*. A field null, defined as $E_z = 0$ (or $\alpha_0 = 0$), can exist at a finite distance from the release point or at infinity. Except in zero gravity, a particle can never quite reach a field null. Marginal radial stability exists when the second radial derivative of E_z^2 goes to zero, a condition expressed by $\alpha_1^2 = 4\alpha_0\alpha_2$. When a particle reaches such a point, it drifts away in the radial direction and is lost.

The locus of levitation is very important in electrode design for passive levitators because it determines the usable dynamic range as well as the sensitivity of the levitator. Figures 3.7a, b, and c depict several practical electrode geometries for passive dielectrophoretic levitation that have been used in experimental applications. In all these structures, the higher field limit of the levitation locus is defined by a release point. Figures 3.7a and b illustrate two examples of electrode structures exhibiting field nulls on the axis. By experimental trial, it has been found that the truncated cylinder geometry of Figure 3.7a offers a good compromise between sensitivity and dynamic range (Jones and McCarthy, 1981). The circular aperture geometry, shown in Figure 3.7c, is distinguished from the other two by the fact that it features a point of marginal radial stability (Jones, 1981).

C.4 Frequency range constraints

Typically, the condition $\text{Re}[\underline{K}(\omega)] < 0$ imposes serious frequency constraints on passive levitation of lossy particles. To illustrate, consider the frequency-dependent behavior of a homogeneous dielectric sphere with permittivity ε_2 and ohmic conductivity σ_2 in a suspension medium with permittivity ε_1 and conductivity σ_1. If there does exist a critical frequency ω_c dividing regions of negative and positive dielectrophoresis, it will be defined by the condition $\text{Re}[\underline{K}] = 0$. Using Equation (3.6), the marginal condition may be written

$$\frac{\varepsilon_2 - \varepsilon_1}{\varepsilon_2 + 2\varepsilon_1} = \frac{3(\varepsilon_2\sigma_1 - \varepsilon_1\sigma_2)}{\tau_{MW}(\sigma_2 + 2\sigma_1)^2(1 + \omega_c^2\tau_{MW}^2)} \tag{3.40}$$

Solving for ω_c, we have

$$\omega_c = \sqrt{\frac{(\sigma_1 - \sigma_2)(\sigma_2 + 2\sigma_1)}{(\varepsilon_2 - \varepsilon_1)(\varepsilon_2 + 2\varepsilon_1)}} \tag{3.41}$$

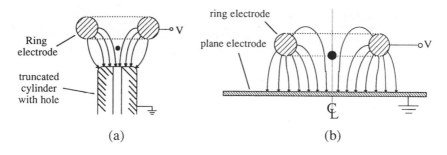

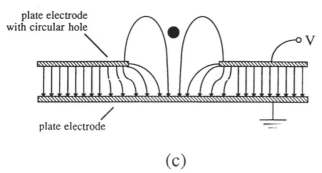

Fig. 3.7 Half sectional views of various axisymmetric electrode structures for passive dielectrophoretic levitators.

(a) The ring-truncated cylinder geometry.

(b) The ring-plane geometry.

(c) The circular aperture geometry.

Note that ω_c is defined only when $(\sigma_1 - \sigma_2)/(\varepsilon_2 - \varepsilon_1) > 0$. Four distinct cases based on the relative values of permittivity and conductivity are shown in Table 3.1. For example, if $\varepsilon_2 > \varepsilon_1$ and $\sigma_2 < \sigma_1$ so that the condition of negative DEP is met at low frequencies but not at high frequencies, a particle can be levitated passively only when $\omega < \omega_c$. If $\varepsilon_2 < \varepsilon_1$ and $\sigma_2 > \sigma_1$, then a particle can be levitated passively only when $\omega > \omega_c$. In the other two distinct cases, ω_c is undefined; passive DEP levitation is either allowed at all frequencies or not allowed at all.

C.5 *Applications of passive DEP levitation*

Passive dielectrophoretic levitation was reported at least as early as 1969 by Veas and Schaffer (1969), who levitated small droplets of an insulating dielectric liquid, immiscible with and suspended in another liquid. The effect has been rediscovered several times since (Parmar and Jalaluddin, 1973; Jones and Bliss,

Table 3.1 *Regimes of possible passive levitation of a homogeneous dielectric particle with ohmic conductivity in a similar dielectric fluid.*

	$\sigma_2 > \sigma_1$	$\sigma_2 < \sigma_1$
$\varepsilon_2 > \varepsilon_1$	Passive levitation is not permitted: ω_c undefined	Passive levitation only at low frequencies: $\omega < \omega_c$
$\varepsilon_2 < \varepsilon_1$	Passive levitation only at high frequencies: $\omega > \omega_c$	Passive levitation at all frequencies: ω_c undefined

1977). Passive DEP levitation has the advantage of simplicity; the particle is held by the nonuniform electrostatic field in static equilibrium, and no particle position detection or feedback control is required. On the other hand, the restriction to particles with negative dielectrophoresis is severe because most interesting particle and biological cell studies must be conducted in frequency ranges where positive DEP reigns. Despite this limitation, some interesting applications have been reported. Veas and Schaffer used passive DEP levitation of dielectric droplets as a simple working model for a new magnetic levitation scheme for handling small quantities of molten metals in metallurgical applications (1969). Kallio and Jones (1980) employed passive DEP levitation to measure the dielectric constant of small solid particles and glass microballoons in the 100 to1000-μm-diameter range. They also measured the dielectric constants of liquids by levitating gas bubbles. Bahaj and Bailey (1979) employed this technique using somewhat smaller particles (diameter ~50 μm) suspended in water, using smaller electrodes and much lower voltages to take advantage of the scaling law, Equation (3.3). They were first to report broadband DEP spectra obtained using levitation. Quite recently, planar quadrupole electrodes fabricated photolithographically on a substrate have achieved passive levitation of biological cells at high frequencies (Washizu and Nakada, 1991; Huang and Pethig, 1991; Fuhr et al., 1992).

D. Feedback-controlled DEP levitation (+DEP)

Figure 3.8 shows a simple electrode geometry used to achieve feedback-controlled DEP levitation. Note that to achieve the condition of Equation (3.37b), the axisymmetric electrodes are shaped to focus the electrostatic field on the axis so that any particle with Re[$\underline{K}$] > 0 is automatically centered. Axial stabilization must be achieved by sensing upward or downward displacements of the particle and using a servo loop to correct the voltage applied to the electrodes.

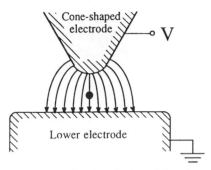

Fig. 3.8 Feedback levitation electrode structure – the cone and plane geometry. (From T. N. Tombs and T. B. Jones, "Effect of moisture on the dielectrophoretic spectra of glass spheres," *IEEE Transactions on Industry Applications*, vol. 29, pp. 281–285, © 1993 IEEE.)

We may construct a simple model for feedback-controlled dielectrophoretic levitation using the dynamic equations of Section 3.4B. As previously shown, any equilibria of the electrostatic field are inherently unstable if $\text{Re}[\underline{K}] > 0$. If we impose the design decision that radial stabilization is to be achieved by passive means and that vertical stabilization is the task of the servo system, then certain constraints are placed upon the cylindrically symmetric electrodes. Imposition of the inequality

$$\alpha_1^2 < 4\alpha_0\alpha_2 \qquad (3.42)$$

in Equation (3.36b) guarantees radial stabilization; however, this same condition also guarantees that any equilibrium will be unstable to axial displacements unless feedback control is employed to adjust the voltage $v'(t)$. The simplest possible realization of this feedback is proportional control, with v' linearly related to the positional displacement $z'(t)$.

$$v'(t) = -GV_0z'(t) \qquad (3.43)$$

where G is an adjustable gain factor. Using Equation (3.43) in Equation (3.36a), the condition for axial stabilization becomes

$$G > \alpha_1/2\alpha_0 + \alpha_2/\alpha_1 \qquad (3.44)$$

Note that the inequality (3.44) has no influence whatsoever upon radial stabilization.

The first demonstration of feedback-controlled DEP levitation was that of Kaler and Pohl (1983), who suspended individual biological cells in a nonuniform electric field by watching the cells and manually adjusting the voltage to maintain them at fixed position. More recently, levitation of metallic spheres (Jones and Kraybill, 1986), chains of metallic spheres (Tombs and Jones, 1991), and plant protoplasts (Kaler and Jones, 1990) using a servo system controlled by a digital computer has been reported. Feedback-controlled DEP levitation of particles exhibiting positive DEP is a breakthrough because most of the particles of interest – biological cells included – exhibit a positive dielectrophoretic effect in typical suspension media at typical frequencies. Computer control of the DEP levitator improves measurement precision and, furthermore, permits automation of the tedious procedure of taking broadband DEP spectral data.

Several basic types of measurements have been conducted upon particles using feedback-controlled dielectrophoretic levitation. Precision measurements performed upon chains of metallic spheres have confirmed theoretical predictions about the influence of chain length on enhancement of the dipole moment (Tombs and Jones, 1991). These measurements are discussed at length in Chapter 6, which covers the subject of particle chaining. Feedback-controlled DEP levitation of particles exhibiting frequency-dependent DEP spectra has been reported by several laboratories. For example, Kaler and colleagues (1990; 1992) measured the membrane capacitance of plant protoplasts and mammalian ligament cells, while Tombs and Jones (1993) investigated the effect of surface moisture on semi-insulating glass beads.

E. Discussion of DEP levitation technique

The *method of fixed position* is the basis of all precision measurements made with the DEP levitator. It is often difficult to obtain an analytical expression or a numerical solution for the nonuniform field $E_z(z)$ of levitator electrodes. Fortunately, there is usually no need for such detailed knowledge of the field, as long as all measurements are made with the particle or particle chain at the same fixed position. One first calibrates the electrodes by levitating a well-characterized particle at the fixed position $z = z_0$. For passive levitation ($\mathrm{Re}[\underline{K}] < 0$), it is often convenient to use a gas bubble, in which case $\varepsilon_2 \approx \varepsilon_o$, the permittivity of free space. For feedback-controlled levitation, any particle such that $\mathrm{Re}[\underline{K}] \approx 1.0$ (for example, a conductive sphere) may be used. All subsequent measurements are then made with reference to this calibration. Note that independent determination of the dielectric properties of the suspension liquid, as well as the mass densities of both the liquid and the particle, are required.

For measurements on particles (and chains) exhibiting dielectric dispersion, the position is maintained by the feedback controller while the frequency is varied. What results is a *dielectrophoretic spectrum*, consisting of voltage versus frequency data: $V_0(\omega)$. If position is fixed, then $\mathrm{Re}[\underline{K}(\omega)] \propto V_0^{-2}$, so that, by plotting the inverse square of the levitation voltage, the frequency dependence of $\mathrm{Re}[\underline{K}]$ is revealed. Note that absolute measurement of $\mathrm{Re}[\underline{K}]$ is possible only if independent calculation of α_0 and α_1 has been performed, or if the electrodes have been calibrated.

Some of the principal factors influencing the accuracy of dielectrophoretic measurements obtained using levitation are summarized below.

E.1 Positional precision

The method of fixed position provides accurate results for a particle only if its position is precisely controlled. Positional control accuracy is strictly limited by the overall optical resolution of the system used to detect the particle position. This resolution is a function of the magnification of the microscope and the inherent spatial resolution of the diode array, CCD chip, video tube, etc.

E.2 RMS voltage measurement precision

The time-average DEP force depends upon the rms magnitude of the electric field in the levitator. A broadband DEP levitator system requires a true-rms AC voltmeter that is accurate over the full range of measurement frequencies.

E.3 Uncertainty in densities of particles and fluids

A DEP levitator is usually calibrated with a metallic or dielectric sphere in a known liquid. The liquid and particle mass densities must be known, because equilibria in a gravitational field are influenced by them. Liquid densities are usually easy to measure independently, but determining the effective density of a small particle is somewhat more difficult. One way to determine the buoyant density of a particle, that is, $\rho_2 - \rho_1$, is to measure the settling velocity (Kaler and Jones, 1990).

E.4 Nonspherical particles

The dielectrophoretic force exerted on a particle depends on its shape. If a particle's shape is not accurately known, then characterization of this particle using DEP levitation will be adversely affected. Refer to Chapter 5, which discusses the orientational behavior of lossy dielectric ellipsoids.

E.5 Sensitivity of Clausius–Mossotti function

The most important limitation on DEP levitation as a measurement tool is that it measures the effective dielectric constant of a particle ε_2' only indirectly. In fact, the real part of the complex Clausius–Mossotti function $\underline{K}(\varepsilon_2', \varepsilon_1)$, defined by Equation (2.34) is the directly measured quantity. Re[$\underline{K}$] of course, contains all the information of interest; however, its sensitivity to ε_2' is rather poor when either $|\varepsilon_2'| >> |\varepsilon_1|$ or $|\varepsilon_2'| << |\varepsilon_1|$ (Kallio and Jones, 1980). Foster's analysis of the sensitivity functions of bio-particle measurements using a symbolic manipulator program shows that dielectrophoresis is better suited to measurement of effective conductivities than to the accurate determination of permittivities (Foster et al., 1992).

E.6 Size-dependent effects

Higher-order multipolar contributions can also influence the accuracy of measurements with a dielectrophoretic levitator (Jones and Loomans, 1983). As illustrated by Equation (2.43), the significance of these higher-order terms relative to the dipole force depends upon particle size. In general, higher-order corrections become important when the particle diameter approaches the length scale of the electric field inhomogeneities (Jones, 1985).

E.7 Other disadvantages and limitations

As a practical matter, Brownian motion prevents particles smaller than a few microns in diameter from being held stationary. At the other extreme, particles larger than approximately 500 μm require large electrodes and excessive voltages. Joule heating of suspension media presents a serious problem in certain circumstances, particularly for biological cells levitated in aqueous media. The levels of ionic concentration required to maintain the viability of mammalian cells can lead to excessive joule heating of the medium. Such heating causes fluid convection, which will seriously degrade levitator performance.

Another problem with DEP levitation stems from the requirement of distinctly different electrode structures for passive and active levitators. Implementation of feedback cannot transform the passive electrode structures shown in Figures 3.7a, b, and c into a feedback-controlled levitator. Likewise, the electrodes of Figure 3.8 cannot be used to levitate a particle when Re[$\underline{K}$] < 0. This limitation leads to serious inconvenience for particles exhibiting +DEP and −DEP in different regions of the frequency spectrum. An excellent solution to this problem is to excite the electrodes of Figure 3.8 with dual frequency voltage, exploring regions of negative dielectrophoresis with the low-frequency signal while maintaining particle position with feedback control of the high-frequency signal (Kaler et al., 1992).

3.5 Magnetophoresis

Magnetizable particles experience a force in a nonuniform magnetic field. This phenomenon, sometimes called magnetophoresis (MAP), has been exploited in a variety of industrial and commercial processes for separation and beneficiation of solids suspended in liquids (Andres, 1976). Much of the foregoing analysis and interpretation of dielectrophoresis carries over to the phenomenology of magnetic particles in a nonuniform field; however, most magnetic particles are ferromagnetic and exhibit strong nonlinearity, limiting somewhat the usefulness of the DEP $\leftrightarrow$ MAP analogy. Furthermore, conductive particles (magnetic or not) in a nonuniform AC magnetic field can exhibit an effectively diamagnetic response because of induced eddy currents. Eddy currents tend to exclude the time-varying field from the particle, causing the particle to be repelled from regions of high magnetic field intensity. The DEP $\leftrightarrow$ MAP analogy breaks down here as well because the magnetic field inside the conductive particle is not curl-free. Still further, superconducting particles, which become effectively diamagnetic in DC magnetic fields, exhibit behavior quite distinct in certain respects from any dielectrophoretic phenomenon.

A. Theory

To achieve the degree of generality necessary for consideration of magnetic particles, we must reconsider the dipole identification problem worked in Section 2.2B. Imagine a homogeneous sphere with radius R and net magnetic polarization $\overline{M}_2$ that is suspended in a magnetically linear fluid of permeability μ_1 and subjected to an almost uniform magnetic intensity $\overline{H}_0$.[5] Recall the defining relationships between $\overline{B}$ (flux density), $\overline{M}$ (volume magnetization), and $\overline{H}$ (magnetic intensity) (Stratton, 1941).

$$\overline{B} = \mu_0(\overline{H} + \overline{M}) \tag{3.45}$$

where $\mu_0 = 4\pi \cdot 10^{-7}$ H/m is the permeability of free space. Equation (3.45) is completely general; the volume magnetization $\overline{M}$ includes permanent magnetization plus any linear or nonlinear function of $\overline{H}$. Such a general treatment allows consideration of any paramagnetic, diamagnetic, or ferromagnetic particle. For present purposes, we assume that $\overline{M}_2 \parallel \overline{H}_0$, where $\overline{H}_0$ is the externally imposed magnetic field intensity vector. Also for the present, we assume that no electric current flows anywhere. Thus, $\nabla \times \overline{H} = 0$ everywhere, and the magnetostatics problem may be solved using a scalar potential Ψ, defined by $\overline{H} = -\nabla \Psi$.

The assumed solutions for Ψ_1 and Ψ_2 outside and inside the sphere, respectively, have forms identical to Equations (2.8a,b).

$$\Psi_1(r,\theta) = -H_0 r \cos\theta + \frac{X \cos\theta}{r^2}, \, r > R \tag{3.46a}$$

---

[5] The analysis here is based upon the analogous problem of a dielectric sphere with a permanent electric dipole moment (Fröhlich, 1958).

and

$$\Psi_2(r,\theta) = -Y\,r\cos\theta,\, r > R \tag{3.46b}$$

where X and Y are constants to be determined by the two boundary conditions at the particle surface. First, the magnetostatic potential must be continuous across the boundary between the particle and the fluid.

$$\Psi_1(r=R,\theta) = \Psi_2(r=R,\theta) \tag{3.47}$$

Second, the magnetic flux density must be continuous across the particle–fluid interface.

$$\mu_1 H_{r1} = \mu_0(H_{r2} + M_{r2}),\, r = R \tag{3.48}$$

where $H_{r1} = -\partial \Psi_1/\partial r$ and $H_{r2} = -\partial \Psi_2/\partial r$ are the normal magnetic field intensity components in the fluid and the sphere, respectively. Using Equations (3.47) and (3.48) with the assumed solutions, the coefficients X and Y may be determined.

$$X = \frac{\mu_0 - \mu_1}{\mu_0 + 2\mu_1} R^3 H_0 + \frac{\mu_0 R^3 M_2}{\mu_0 + 2\mu_1} \tag{3.49a}$$

$$Y = \frac{3\mu_1}{\mu_0 + 2\mu_1} H_0 - \frac{\mu_0}{\mu_0 + 2\mu_1} M_2 \tag{3.49b}$$

Note that Y is the magnitude of the uniform magnetic field H_2 inside the sphere. We may determine the effective magnetic dipole moment $\overline{m}_{\text{eff}}$ by comparing the dipole term in the assumed solution Equation (3.46a) to

$$\Psi_{\text{dipole}} = \frac{m_{\text{eff}} \cos\theta}{4\pi r} \tag{3.50}$$

The result is

$$\overline{m}_{\text{eff}} = 4\pi X = 4\pi R^3 \left[\frac{\mu_0 - \mu_1}{\mu_0 + 2\mu_1} \overline{H}_0 + \frac{\mu_0}{\mu_0 + 2\mu_1} \overline{M}_2 \right] \tag{3.51}$$

Remember our assumption that $\overline{M}_2 \parallel \overline{H}_0$. The first term is the contribution to the net moment of the fluid (having linear permeability μ_1) displaced by the particle, while the second is due to any magnetization of the particle itself. In the limit of a magnetized particle in vacuum, that is, $\mu_1 = \mu_0$, Equation (3.51) agrees with other results (Smythe, 1968, Sec. 9.10).

B. Magnetically linear particle

For a magnetically linear particle of radius R and permeability μ_2, we have

$$\overline{M}_2 = \chi_2 \overline{H}_2 \tag{3.52}$$

where $\chi_2 \equiv \mu_2/\mu_0 - 1$ is the susceptibility of the particle. In this special case, the effective moment vector takes a form very similar to that for the effective dipole moment of dielectric sphere.

$$\overline{m}_{eff} = 4\pi R^3 K(\mu_2,\mu_1)\overline{H} \tag{3.53}$$

$$K(\mu_2,\mu_1) = \frac{\mu_2 - \mu_1}{\mu_2 + 2\mu_1} \tag{3.54}$$

The Clausius–Mossotti function retains its original form, with permeabilities μ_1 and μ_2 replacing permittivities ε_1 and ε_2. One distinction between the definitions for p_{eff} and m_{eff} in Equations (2.12) and (3.53), respectively, is that the factor μ_1 is missing from the latter. The reason for this minor discrepancy is more historical than fundamental and has to do with differences in the ways that the field theories of dielectric and magnetic materials were originally formulated over a century ago.

The missing μ_1 term reappears in the expression for the force exerted by a nonuniform magnetic field on a magnetic dipole.

$$\overline{F}_{dipole} = \mu_1 \overline{m}_{eff} \cdot \nabla \overline{H}_0 \tag{3.55}$$

Combining this expression with that for the effective moment, the magnetophoretic force for a magnetizable spherical particle in a nonuniform magnetic field may be written as

$$\overline{F}_{MAP} = 2\pi \mu_1 R^3 K(\mu_2,\mu_1)\nabla H_0^2 \tag{3.56}$$

Note that the MAP force expression has a form analogous to Equation (3.1). Refer to Tables A.1 and A.2 in Appendix A for a comprehensive summary of the analogy between DEP and MAP.

Based on the completely analogous forms of Equations (3.1) and (3.56), we may establish and summarize the fundamental phenomenology of magnetophoresis by reference to the list provided in Section 3.2B for dielectrophoresis.

(i) F_{MAP} is proportional to particle volume.

(ii) F_{MAP} is proportional to the permeability of the suspension medium μ_1.

(iii)   The MAP force is directed along the gradient of the magnetic field intensity ∇H_0^2.

(iv)   The MAP force depends upon the magnitude and sign of the Clausius–Mossotti function K, but with μ_1 and μ_2 replacing ε_1 and ε_2.

Statements (iii) and (iv) imply that regions of higher magnetic field intensity serve to attract or repel a magnetizable particle, depending upon the relative magnitudes of μ_1 and μ_2. Consistent with the definitions of positive and negative dielectrophoresis, we define:

Positive magnetophoresis: $K > 0$ (or $\mu_2 > \mu_1$). Particles are attracted to magnetic field intensity maxima and repelled from minima.

Negative magnetophoresis: $K < 0$ (or $\mu_2 < \mu_1$). Particles are attracted to magnetic field intensity minima and repelled from maxima.

Assuming that the imposed magnetic field is divergence- and curl-free, we may show that

$$\nabla^2 H_0^2 \geq 0 \tag{3.57}$$

This result, analogous to Equation (3.22), means that a magnetic field can have no local intensity maxima but can have local minima, including zeros. The implications for levitation of magnetic particles in divergence- and curl-free magnetic fields may be anticipated: passive levitation is possible for particles exhibiting negative magnetophoresis ($K < 0$), and feedback-controlled levitation is required for particles with positive magnetophoresis ($K > 0$).

C. Nonlinear magnetic media

Unlike most dielectrics, magnetizable materials exhibit strongly nonlinear behavior such as paramagnetism and ferromagnetism, which seldom can be ignored in modeling the electromechanics of magnetic particles. Though magnetization phenomena are very complex, it is adequate, for the purposes of describing the forces on magnetizable particles, to group nonlinear magnetic materials into two classifications, namely, "soft" and "hard" materials. Representative magnetization curves of M versus H are shown for these two types of materials in Figures 3.9a and b. The principal manifestation of nonlinearity in magnetically soft materials is *saturation*, which limits the magnitude of the magnetization vector to a finite value M_{sat}. Saturation, an attribute of orientational polarization, occurs when all the domains become aligned. The influence of saturation upon the effective moment of a particle should be evident.

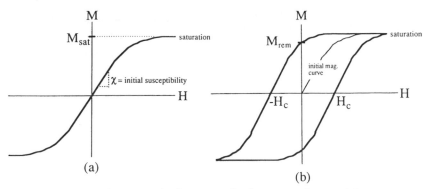

Fig. 3.9 Representative magnetization curves for ferromagnetic materials.
(a) Magnetically soft material showing saturation.
(b) Magnetically hard material showing hysteresis loop.

Magnetically hard materials exhibit saturation, but just as important, they require significant energy to realign their domains once the material is first magnetized. The phenomenon, represented by the *M–H* curve in Figure 3.9b, is called *hysteresis*. The shape of the hysteresis curve is dependent upon the *hardness* of the material, plus the magnitude and rate of change of the applied magnetic field. Hysteresis makes possible permanent magnetism, but it also represents an energy loss mechanism producing heat. Hysteretic behavior can be important in the mechanics of magnetic particles because it creates a phase lag between the imposed magnetic field $\bar{H}_0$ and the effective magnetic moment $\bar{m}_{\text{eff}}$. This phase lag is important in particle rotation.

It is a practical decision to restrict the coverage of nonlinear magnetophoresis within this chapter to two important cases: (i) ferromagnetic particles in a linear magnetizable medium, and (ii) nonmagnetic particles in a magnetizable liquid, such as a ferrofluid. In both cases, the magnetic field is assumed to be DC and hysteresis is ignored.

Consider a spherical ferromagnetic particle immersed in a linear magnetizable fluid of permeability μ_1. Assume that the particle magnetization is some nonlinear function of the field, so that $\bar{M}_2(\bar{H}_2) \parallel \bar{H}_0$, where $\bar{H}_2$ is the uniform field within the particle. Then, Equations (3.51) and (3.55) may be used to obtain the MAP force.

$$\bar{F}_{\text{MAP}} \approx 2\pi\mu_1 R^3 \left[\frac{\mu_0 - \mu_1}{\mu_0 + 2\mu_1} \nabla H_0^2 + \frac{2\mu_0}{\mu_0 + 2\mu_1} \bar{M}_2(\bar{H}_2) \cdot \nabla \bar{H}_0 \right] \tag{3.58}$$

To use this expression, $\bar{H}_2$ must be obtained in terms of $\bar{H}_0$. If the particle is magnetically soft and the magnetic field is strong, then the particle will saturate so that $M_2 \rightarrow M_{\text{sat}}$.

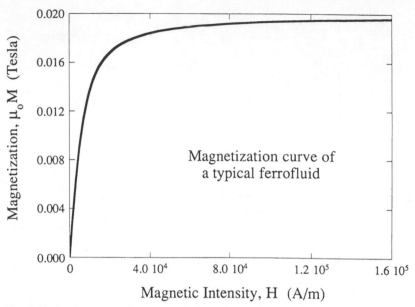

Fig. 3.10 Typical magnetization curve $M(H)$ for a ferromagnetic fluid.

Nonlinear magnetic properties such as saturation and hysteresis have no significant qualitative effect on the nature of magnetophoresis. For example, saturation influences only the magnitude of $\bar{F}_{MAP}$ by limiting the effective moment; particles will still be attracted to regions of higher or lower magnetic field intensity depending on whether they are, respectively, more or less magnetizable than the suspension medium. Hysteresis, which creates a phase difference between the magnetic field and the induced moment, has more influence on rotation and is discussed in Section 4.5B. The problem of a particle with crystalline anisotropy is examined in Section 5.6A.

Ferrofluids are stable colloidal suspensions of subdomain-sized magnetite particles in a liquid vehicle. These magnetic liquids offer potential application in recycling and other materials separation processes where a nonuniform magnetic field is used. Ferrofluids exhibit weak superparamagnetic behavior typified by the *M–H* curve shown in Figure 3.10, from which it should be evident that the saturation magnetization of ferrofluids is very low compared to that of ferromagnetic materials. Of primary interest here is the observable force on a nonmagnetic object, referred to as the *magnetofluid buoyancy effect* of the first kind (Rosensweig, 1966). The magnetic nonlinearity of ferrofluids creates difficulties, making unambiguous identification of an effective magnetic moment for an immersed particle difficult or impossible. Therefore, we use a Maxwell stress tensor formulation to predict the magnetic force $\bar{F}_{ff}$ on nonmagnetic immersed particles (Rosensweig, 1979a).

$$\bar{F}_{\text{ff}} = -\oint_{\Sigma} \left\{ \mu_0 \frac{M_n^2}{2} + \mu_0 \int_0^{H_0} M(H')dH' \right\} \hat{n} dS \tag{3.59}$$

where Σ is the surface that just encloses the object, H_0 is the magnitude of the non-uniform magnetic field, $\hat{n}$ is the outward-directed unit normal vector at the surface, and $M_n = \hat{n} \cdot \bar{M}(\bar{H})$ is the normal component of the magnetization in the ferrofluid adjacent to the surface Σ enclosing the particle. The influence of the field gradient becomes evident only when one realizes that the surface integral of Equation (3.59) goes to zero unless $\bar{H}_0$ is nonuniform.

In a very weak magnetic field, we may linearize the M–H curve to define an effective value of permeability μ_1, in which case Equation (3.56) can be used by setting $\mu_2 = \mu_0$. Unfortunately, this limit is of little practical use in separation technology because of the low value of initial permeability. To achieve a substantial force, magnetic fields strong enough to saturate the ferrofluid are required, so that $|\bar{M}| \approx M_{\text{sat}} \ll |\bar{H}_0|$. In this case, Equation (3.59) may be replaced by a far simpler result.

$$\bar{F}_{\text{ff}} \approx -\mu_0 V M_{\text{sat}} \nabla H_0 \tag{3.60}$$

where V is the volume of the particle. Particle shape has no influence on this force expression because demagnetization is not significant when $M_{\text{sat}} \ll |\bar{H}_0|$. Only in this saturated limit is it possible to extract, somewhat artificially, an effective dipole moment value: $\bar{m}_{\text{eff}} \approx -V\bar{M}_{\text{sat}}$.

D. Magnetophoresis with eddy current induction

When a time-varying magnetic field is applied to an electrically conductive magnetizable particle, eddy currents are induced, which tend to oppose penetration of the field into the particle. At sufficiently high frequencies, these eddy currents emulate diamagnetism and strongly influence the effective magnetic moment. The formal analogy between dielectric and magnetic particles breaks down completely for conductive magnetizable particles because, within the particle, $\nabla \times \bar{H} \neq 0$. Thus, a solution must be sought using the magnetic vector potential.

Imagine a small magnetic dipole of magnitude $\bar{m}_{\text{eff}}$ in a linear medium of permeability μ_1. From Smythe (1968), the magnetic vector potential $\bar{A}$ at position $\bar{r}$ may be written as

$$\bar{A} = \frac{\mu_1 \bar{m}_{\text{eff}} \times \bar{r}}{4\pi r^3} \tag{3.61}$$

where $\nabla \times \bar{A} \equiv \bar{B}$, $\bar{B}$ is the magnetic flux density, and $r = |\bar{r}|$. There is no loss of generality if we align the moment parallel with the z axis; that is, $\overline{m}_{eff} = m_{eff}\hat{z}$. Then Equation (3.61) becomes

$$\overline{A} = \mu_1 \frac{m_{eff} \sin\theta}{4\pi r^2}\hat{\phi} \tag{3.62}$$

where $\hat{\phi}$ is the azimuthal unit vector in spherical coordinates. In the identification of the magnetic dipole moment when eddy currents are present, Equation (3.62) serves a function similar to Equation (3.50). To proceed, the boundary value problem is solved in terms of the vector potential and then the dipole term of the vector potential outside the particle is compared to Equation (3.62).

Smythe (1968) published a solution to the problem of a magnetizable, conductive sphere of permeability μ_2 and conductivity σ_2 immersed in a non-conductive fluid of permeability μ_1 and subjected to a uniform, linearly polarized, magnetic field $\overline{H}(t) = \hat{z}H_0 \cos\omega t$. From an examination of his solution for $\overline{A}$ outside the particle, we can identify the effective moment of the sphere.

$$\overline{m}_{eff} = 4\pi R^3 \underline{L}\,\overline{H}_0\hat{z} \tag{3.63}$$

where $\underline{L}(\omega)$ is a frequency-dependent magnetization coefficient that takes the place of $\underline{K}(\omega)$. Note that $\underline{L} = R^3\underline{D}/2$ and that $\underline{D}$, which depends on μ_1, μ_2, σ_2 and ω, is defined as follows (Smythe, 1968).

$$\underline{D} = \frac{(2\mu_2 + \mu_1)vI_{-1/2} - [\mu_1(1+v^2) + 2\mu_2]\,I_{+1/2}}{(\mu_2 - \mu_1)vI_{-1/2} + [\mu_1(1+v^2) - \mu_2]\,I_{+1/2}} \tag{3.64}$$

with

$$I_{\pm 1/2} \equiv I_{\pm 1/2}(v) \quad \text{and} \quad v \equiv \sqrt{j\sigma_2\mu_2\omega}R \tag{3.65}$$

The quantities $I_{\pm n/2}$ are half-integer order, modified Bessel functions of the first kind. Unlike the Clausius–Mossotti function $\underline{K}(\omega)$ for a homogeneous dielectric particle, the coefficient $\underline{L}(\omega)$ does not represent a single relaxation process and so, presumably, cannot be manipulated into the form of Equation (2.31).

Given $\overline{m}_{eff}$, we can calculate the dipole contribution to the magnetophoretic force. For the dipole approximation to be valid, any nonuniformity of the imposed magnetic field must be modest. According to the effective dipole method, the time-average force is

$$\langle \overline{F}_{MAP}(t) \rangle = \frac{\mu_1}{2} \text{Re}\,[(\overline{m}_{eff} \bullet \nabla)\,\overline{H}^*] \tag{3.66}$$

Now, Equations (3.63) and (3.66) may be combined to obtain an expression for the magnetophoretic force in terms of the real part of $\underline{L}$.

$$\langle \bar{F}_{MAP}(t)\rangle = 2\pi\mu_1 R^3 \text{Re}[\underline{L}]\nabla H_0^2 \tag{3.67}$$

A simplified expression for $\text{Re}[\underline{L}]$ may be obtained by invoking certain Bessel function identities (Holmes, 1978).

$$\text{Re}[\underline{L}] = \frac{3\mu_R}{2}\frac{(\mu_R - 1)B^2 + R_m^2(\sinh 2R_m - \sin 2R_m)A}{(\mu_R - 1)^2 B^2 + R_m^2 A^2} - \frac{1}{2} \tag{3.68}$$

where $\mu_R = \mu_2/\mu_1$ is the relative permeability and $R_m = R\sqrt{\omega\mu_2\sigma_2/2}$, a dimensionless modulus called the magnetic Reynolds number, compares the particle radius to the skin depth (Woodson and Melcher, 1968).[6] Other definitions used in (3.68) are as follows:

$$A = 2R_m(\cosh 2R_m - \cos 2R_m) + (\mu_R - 1)(\sinh 2R_m - \sin 2R_m) \tag{3.69a}$$

$$B = \cosh 2R_m - \cos 2R_m - R_m(\sinh 2R_m + \sin 2R_m) \tag{3.69b}$$

Depending on the sign of $\text{Re}[\underline{L}]$, the magnetophoretic effect is positive or negative (as defined previously in Section 3.5B). It may be shown readily that $-0.5 \leq \text{Re}[\underline{L}] \leq 1.0$. The frequency dependence of $\text{Re}[\underline{L}]$ is shown as a function of the magnetic Reynolds number R_m ($\propto \sqrt{\omega}$) in Figure 3.11 for several values of relative permeability μ_R. Comparison of these curves to Figures 3.2a, and b reveals marked differences between the frequency dependence of $\text{Re}[\underline{K}]$ and $\text{Re}[\underline{L}]$.

The low-frequency ($R_m \ll 1$) and high-frequency ($R_m \gg 1$) limits are, respectively

$$\lim_{R_m \to 0}[\underline{L}] = \frac{\mu_2 - \mu_1}{\mu_2 + 2\mu_1} \tag{3.70a}$$

$$\lim_{R_m \to \infty}[\underline{L}] = -0.5 \tag{3.70b}$$

As expected, the low-frequency limit, where no significant eddy current flow occurs, reduces to the case of a sphere with permeability μ_2 in a fluid of permeability μ_1, identical to Equation (3.53). Also as expected, the high-frequency

---

[6] Equation (3.68) may be reconciled with Holmes's result (1978) by using the following definition: $G(R_m,\mu_R) \equiv -2\text{Re}[\underline{L}]$.

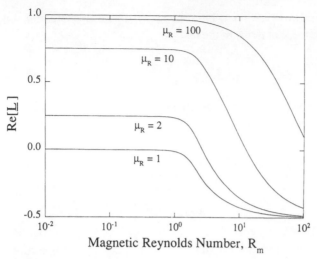

Fig. 3.11 Plots of Re[$\underline{L}(\omega)$] versus the magnetic Reynolds number R_m for various values of relative magnetization: $\mu_R = 1$, 2, 10, and 100.

case reduces to the perfect diamagnetic limit with no magnetic field penetrating the particle.

For a nonmagnetic conductive spherical particle in air, $\mu_2 = \mu_1 = \mu_0$ and the expression for Re[$\underline{L}$] can be greatly simplified.

$$\text{Re}\,[\underline{L}] = \frac{3}{8R_m} \frac{\sinh 2R_m - \sin 2R_m}{\sinh^2 R_m + \sin^2 R_m} - \frac{1}{2} \tag{3.71}$$

The DC and high-frequency limits of Equation (3.71) are 0 and -0.5, respectively.

The frequency-dependent diamagnetic effect of the induced eddy currents upon the time-average MAP force $<F_{MAP}(t)>$ is cloaked in the algebraic complexity of Equation (3.68). One simple way to examine the frequency dependence is to identify the marginal condition Re[$\underline{L}(\omega)$] = 0 separating +MAP from −MAP for a conductive, magnetizable particle. Figure 3.12 plots this marginal condition for a sphere in an AC magnetic field with respect to μ_R and R_m. Note how the boundary between positive and negative MAP is defined by $\mu_R \approx 1.0$ at low frequencies and $R_m \propto \mu_R$ at high frequencies.

Combining of Equation (3.67) with the condition on the curl-free imposed magnetic field $\bar{H}_0$ represented by Equation (3.57) leads us to conclude that passive magnetophoretic levitation is possible only when Re[$\underline{L}$] < 0. Consistent with the line of reasoning employed in Section 3.4A, we may state the following theorem regarding passive levitation of conductive, magnetizable particles:

Passive magnetophoretic levitation of a conductive particle is possible only if the particle exhibits a negative magnetophoretic effect, that is, Re[$\underline{L}(\omega)$] < 0.

This theorem has implications for the levitation of metal particles and droplets in magnetic fields for the containerless processing of conductive alloys.

E. Force on superconducting particles

A conventional type I superconducting particle exhibiting the Meissner effect may be modeled as a perfect diamagnetic object. Such a model is adequate as long as the magnetic field intensity does not exceed the critical field strength (or the corresponding critical surface current) of the material. If these limits are not exceeded, we may estimate the magnetic moment $\bar{m}_{eff}$ for a type I superconducting particle by taking the $K = -0.5$ limit of Equation (3.53).

$$\bar{m}_{eff} = -2\pi R^3 \bar{H}_0 \tag{3.72}$$

The magnetophoretic force is then

$$\bar{F}_{MAP} = -\pi\mu_1 R^3 \nabla H_0^2 \tag{3.73}$$

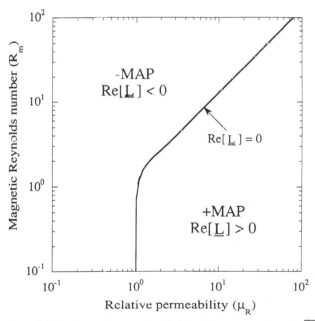

Fig. 3.12 Positive and negative MAP regions versus $R_m = R\sqrt{\omega\mu_2\sigma_2/2}$ and $\mu_R = \mu_2/\mu_1$ for spherical conductive particle of radius R, permeability μ_2, and conductivity σ_2 (from Jones, 1979a).

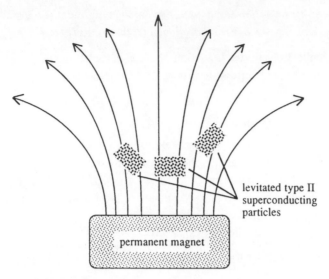

levitated type II
superconducting
particles

permanent magnet

Fig. 3.13 Several orientations of type II superconductive particles levitated passively in the cusped field of a permanent magnet.

Equation (3.73) indicates that superconductive particles will always seek minima of the magnetic field intensity and can be levitated passively in a cusped magnetic field.

Subject to the limitations imposed by the critical field/current condition, Equation (3.73) adequately describes the force on a type I superconducting spherical particle. For type II superconductors, such as the recently discovered high-T_c ceramics, the situation is greatly complicated by flux pinning effects. Flux pinning means that a levitated particle can assume almost any orientation because of flux lines trapped within the particle (Brandt, 1990). The trapped flux creates a restoring force that is effectively conservative for small displacements (Johansen and Bratsberg, 1993) but large displacements alter the equilibrium and dissipate energy. In effect, an infinite number of stable equilibria are possible, as shown in Figure 3.13. Such non-uniqueness is not observed in the case of type I superconductors because, when all flux lines are excluded from the particle, only a single energy-minimizing equilibrium exists.

3.6 Applications of dielectrophoresis and magnetophoresis

A. Biological DEP
The first significant demonstration of biological dielectrophoresis involved the separation of living biological cells (Pohl, 1978). Pohl envisioned both clinical

and laboratory applications of such a technology. Because of rapid progress made over the past decade, biological DEP certainly merits special focus in any discussion of applications. In this section, we review some of the more important advances made since the publication of Pohl's book in 1978.

A.1 Electrofusion

A dramatic success of dielectrophoresis is its use by Zimmermann and his colleagues in electric field–mediated cell fusion (Zimmermann and Vienken, 1982). In this procedure, a suspension containing the two cell types to be fused is introduced into a chamber with an electrode array that creates a nonuniform electric field. The nonuniform field, of typical frequency 100 to 1000 kHz and magnitude ~100 V/cm, collects some fraction of these cells on electrode surfaces where cells of the two types inevitably encounter each other and form chains. When sufficient chains of the desired type have formed, a series of short DC pulses, typically ~10 μsec at ~1 kV/cm, is applied to the electrodes. The strong DC pulses puncture or disrupt the membranes in the region of contact between cells and initiate their merger or fusion into hybrids that, presumably, incorporate desirable attributes of each constituent. Potential applications for this technology are envisioned in the production of hybridomas for antibody production useful in cancer research and treatment.

A.2 Cell manipulation

Washizu has investigated applications of DEP in the control and manipulation of cells and other biological particles, focusing upon development of practical technologies for the automated processing of cells and other biological particles, such as DNA (Washizu, 1990). He has developed a microminiature assembly line for cell fusion called a *fluid-integrated circuit* (FIC), which incorporates photolithographically fabricated electrode and insulator contours with a miniaturized piezoelectric pump (Masuda et al., 1989; Washizu, Nanba, and Masuda, 1990). The capability to transport cells, one at a time, through complex processing protocols using these rather simple electrode configurations has been demonstrated. One example is the unique cell positioning system shown in Figure 3.14a, which assembles cell chains prior to electrofusion. The insulating barrier constricts the electric current to flow through the aperture, thus intensifying the electric field so that two cells introduced on either side of the barrier are attracted to the aperture where they form a cell chain. The cells can then be fused using a strong DC electric field pulse. The cell shift register of Figure 3.14b works on a similar principle; by applying an AC voltage sequentially to the electrodes, single cells are transported along the structure one at a time.

Washizu also reports success in collecting and positioning DNA molecules

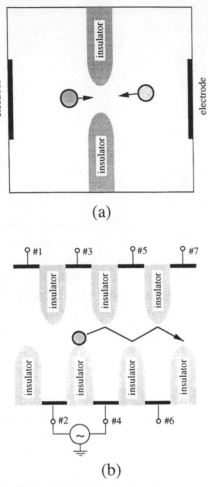

(a)

(b)

Fig. 3.14 Fluid integrated circuit concepts of Washizu (1990) using dielectrophoretic effect for handling and manipulation of biological cells.

(a) Electrode/insulator structure for automated electrofusion system.

(b) Dielectrophoretic cell shift register.

and other biological macromolecules with microfabricated electrodes (Washizu and Kurosawa, 1990; Washizu et al., 1992b; Washizu et al., 1993b). In effect, the dielectrophoretic force is harnessed to collect and position the particles. That the DEP is effective for particles so small (diameter $\sim 5 \cdot 10^{-9}$ m) is somewhat surprising because of the importance of Brownian motion in this size range. Certain aspects of this work relate to particle orientation in an electric field and are discussed in Section 5.7.

A.3 Basic cell studies

One method of investigating cell structure is using DEP measurements to develop models for their dielectric properties. The task is usually posed to the investigator as an inverse problem, where cell structure must be inferred from measured DEP spectra. Quite often, an investigator does not know either how complex the model should be or if the measurement precision is adequate to reveal desired information about cell structure. Despite these difficulties, some important contributions have been made with a variety of different techniques that utilize dielectrophoresis.

Marszalek used a simple vertically aligned set of closely spaced electrode wires with variable-frequency AC voltage applied between them to obtain cell spectra (Marszalek et al. 1989). By releasing cells one at a time and then monitoring them with a microscope as they sank, he observed their response to the two-dimensionally nonuniform electric field. The technique of obtaining data at different frequencies was to record the (inertialess) motion in the horizontal plane for applied voltages at different frequencies. Negative and positive DEP spectra of *Neurospora crassa* (slime) were obtained in this way.

A technique more suited for measuring the DEP spectra of cell populations uses interdigitated electrodes mounted in an optical sample chamber to monitor the frequency-dependent collection of cells (Price et al., 1988; Burt et al., 1989). An advantage of this system is that it is readily automated. An optical detector measures the DEP cell collection by monitoring light transmission through the chamber. To distinguish negative from positive DEP, fairly sophisticated image analysis methods are required (Gascoyne et al., 1992). The technique has been used over the frequency range from 1 Hz to 2 MHz to investigate the effects of certain chemical agents upon the surface charge and membrane conductivity in mammalian cells.

The unique capability of DEP levitation is its capability of measuring single cells and chains. Figure 3.15 illustrates the essential features of a simple computer-controlled DEP levitator system. The particle is illuminated from the side and its shadow is focused on a photodiode array or video camera. Axial displacements of the particle are translated to a digital signal that is converted to analog, amplified, and used to modulate the amplitude of the AC levitation voltage. The technique has been used to measure the membrane capacitance c_m of single-plant protoplasts and (mammalian) ligament cells (Kaler and Jones, 1990). This is accomplished by obtaining the DEP spectra of individual cells and then comparing them to the predicted spectra of the simple spherical shell model of Figure 3.4a. More recently, a dual-frequency method has been demonstrated that permits investigation of the low-frequency region of the dielectrophoretic spectrum

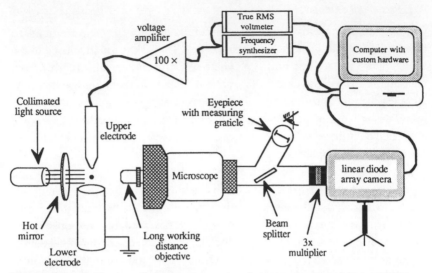

Fig. 3.15 Feedback-controlled levitation system for automated acquisition of DEP spectral data (From T. N. Tombs and T. B. Jones, "Effect of moisture on the dielectrophoretic spectra of glass spheres," *IEEE Transactions on Industry Applications*, vol. 29, pp. 281–285, © 1993 IEEE.)

(<1000 Hz), where negative dielectrophoresis reigns and where poorly understood surface charging effects are evident (Kaler et al., 1992).

B. Microactuators

Recent advances in microfabrication techniques have set the stage for new classes of passive DEP levitator structures that differ dramatically from the axisymmetric electrodes depicted in Figures 3.7 and 3.8. In several research laboratories, planar, salient electrodes have been fabricated on a silicon substrate using photolithographic techniques to create an azimuthally periodic electric field (Washizu and Nakada, 1991; Huang and Pethig, 1991; Fuhr et al., 1992). These new electrode geometries pave the way for micromotors that simultaneously levitate particles and induce rotation. The levitation effect in such structures as the planar quadrupole is due to an induced quadrupolar moment, rather than the dipole (Washizu et al., 1993a). A generalized model for the stability of planar levitators of this type has been proposed (Jones and Washizu, 1994).

Interesting experiments have been reported using planar arrays of electrodes to produce a traveling electric field wave that can transport cells in an assembly line fashion (Batchelder, 1983; Hagedorn et al., 1992). Interdigitated electrode structures were fabricated on glass and on silicon substrates using photolithographic techniques. Very strongly frequency-dependent transport of pollen grains and cellulose spheres has been demonstrated using electric field frequencies in the range of 10^5 to 10^6 Hz.

C. DEP mineralogical separations

Conventional electrostatic methods, which rely upon differences in the electrical charging tendencies of various constituents of crushed ores and minerals, have been used in dry separation and beneficiation processes for many years. Wet DEP separation processes, which use variable frequency, nonuniform AC electric fields, offer an alternative in certain commercially important mineral recovery operations (Benguigui, Shalom, and Lin, 1986; Shalom and Lin, 1988). The dielectrophoretic force is typically much weaker than the coulombic force used in dry electrostatic processes; however, DEP has the advantage of being controllable by the frequency. The approach, analogous to high-gradient magnetic separation, is to establish a highly nonuniform electric field throughout a large volume. This can be done by introducing a matrix of fibrous dielectric material into the space between two electrodes. These fibers create locally strong nonuniformities of the electric field adequate for trapping mineral species in the fiber matrix. Using variable-frequency AC electric fields and properly engineered flow conditions to achieve desired particle residence times, separations based upon both dielectric constant and electrical conductivity are possible. Solid/particulate separation experiments have also been reported using interdigitated electrodes (Benguigui and Lin, 1988).

In most if not all mineral separation processes, particle charging is inevitable. Recognizing this, Goossens and van Biesen (1989) numerically calculated the trajectories of charged dielectric particles in a nonuniform AC electric field. Their results reveal that the oscillatory motions of changed particles induced by the AC coulombic force produce a nonzero time-average force contribution that mimics dielectrophoresis.

D. High-voltage liquid insulation

High-voltage equipment using liquid insulation often fails when gaseous, liquid, or particulate impurities find their way into the liquid. Because of the strong electric fields within the liquid, particles are often subjected to combined coulombic and dielectrophoretic forces. Particles that are conductive or have high dielectric constant are naturally drawn into the regions of the most intense field where they can cause the greatest harm. Some very interesting investigations of the dynamic behavior of solid particles in practical geometries have been reported (Molinari and Viviani, 1978; Bozzo et al., 1985a; Bozzo et al., 1985b).

In related work, Feeley and McGovern (1988) investigated the motion of gas bubbles in a nonuniform electric field.

E. MAP separation technologies

Virtually all practical magnetic separation technologies exploit the magneto-phoretic forces discussed in Section 3.5. Examples include belt and drum devices, high-gradient magnetic separation (HGMS) systems, and magnetohy-drostatic and magnetohydrodynamic separation schemes. Applications cover a broad range of industrial activities (Oberteuffer, 1974): mineral separation (ben-eficiation of clays and iron ores), processing of pulverized coal (desulfurization and cleaning), waste water treatment and purification (steel mills, nuclear reac-tors, steam condensers), particulate emissions (fly ash removal, iron oxide removal), blood separations (red blood cells), and even ocean sediment separa-tions (searching for magnetic monopoles).

HGMS is a promising technology for removal of a wide variety of small, low-susceptibility particulates from water (Oberteuffer, 1973; Watson, 1973). HGMS works by virtue of very strong magnetic field gradients produced throughout a large volume created by loading the separator with fibrous mag-netic material such as steel wool and placing the unit between the poles of a strong magnet. The magnetic matrix creates highly localized field nonuniformi-ties throughout the separator and particles are attracted to the magnetic matrix by these field nonuniformities. HGMS device performance is determined pri-marily by the strength of the magnet. Applications in primary treatment of industrial wastewater have been examined, for example, the removal of iron by-products from wastewater in steel production plants. Another application is the protection of turbine blades in power plants by removal of microscopic scale particles from live steam.

One variant of magnetic separation, *magnetohydrostatic (MHS)*, uses the MAP force in Section 3.5 to achieve separation or beneficiation of larger solids suspended in aqueous suspensions (including solutions of magnetic salts such as $MnCl_2$) (Andres, 1976). Magnetogravimetric separation methods, based upon the differences in magnetic properties and densities of solids, have been demonstrated. It is sometimes possible to formulate magnetic solutions with susceptibilities intermediate between those of two principal mineral species to be separated. Then, as predicted by Equation (3.56), one mineral specie will exhibit a positive and the other a negative MAP effect. In MHS technology, specially shaped magnetic pole pieces produce a magnetic field of specified nonuniformity to achieve continuous beneficiation. One special pole shape, the *isodynamic* pole, produces a uniform force at all elevations, that is, $\partial H_0^2 / \partial z \approx$ constant, where z is the vertical coordinate. Ferrofluids, stable colloidal suspensions of magnetite, have higher susceptibility than magnetic salt solutions and thus provide stronger separation forces, but they are expensive and require efficient recovery/recycling operations to achieve economic feasibility (Khalafalla, 1976).

Ferrography is a specialized application of MAP separation used to examine lubricating oils for signs of abnormal machine wear (Scott et al., 1974). Samples of oil are drawn and then wear particles are magnetogravimetrically separated out using a nonuniform magnetic field. Microscopic evaluation of the numbers, sizes, and shapes of these particles then provides information about the condition inside internal combustion engines.

Another variant of magnetic separation, somewhat related to MHS, is based upon the magnetohydrodynamic (MHD) effect. The net force on a particle in crossed current and magnetic fields is due to differences between the conductivities of the fluid and the particle (Andres, 1976).

F. Magnetic particle levitation

No treatment of particle electromechanics is complete without mention of magnetic levitation. One early application was containerless processing of molten metals and alloys to avoid the inevitable contamination due to crucibles. Another was the free suspension of airfoil designs in wind tunnels without mechanical support. Modern applications for magnetic levitation run from billion-dollar technologies such as magnetically levitated (MAGLEV) trains to esoteric physics experiments such as the quests for gravitational waves and fractional charge.

In recent years, a great variety of interesting applications for both passive and feedback-controlled levitation of magnetic particles has been reported. One such example is a sensitive high-vacuum gauge that uses passively levitated diamagnetic particles (graphite and bismuth, less than 1 mm in size) levitated in cusped magnetic field geometries (Kendall et al., 1987). Hebard (1973) levitated superconducting niobium spheres, also less than 1 mm, in an instrument designed to search for the elusive fractional charge attributed to quarks. The recent discovery of high-temperature ceramic superconductors has provoked renewed interest in passive magnetic levitation (Brandt, 1990; Johansen and Bratsberg, 1993).

Many applications of feedback-controlled levitation of magnetizable particles have been reported over the years, including a device for measuring the viscosity of small sample quantities of liquids (Leyh and Ritter, 1984). According to this method, a small permanent magnet is imbedded inside a 3-mm precision polypropylene sphere. The particle is suspended in the liquid and driven by a low-frequency AC component of the magnetic field to create oscillatory motion. The known phase relationship between force and displacement is then employed to determine the viscosity.

4

Particle rotation

4.1 Introduction

As early as 1892, Arno reported that small particles can be made to spin when placed in a rotating electric field. This rotation is not synchronous; the angular velocity of the particle depends on the electric field magnitude squared. In recognition of significant later contributions made by Born (1920) and Lertes (1920, 1921), this phenomenon has become known as the Born–Lertes effect. Shortly after Arno's work, Weiler (1893) discovered the related phenomenon that small solid particles suspended in liquids can rotate in a static electric field. Though Weiler published first, this effect has come to be known as Quincke rotation (Quincke, 1896). Unlike the Born–Lertes effect, Quincke rotation is a threshold phenomenon; spontaneous rotation occurs once the field strength is increased above some critical value. The rotational axis is always perpendicular to the imposed electric field for both effects. Pickard (1961) reviewed and clarified these phenomena, emphasizing the close relationship of the two effects. This chapter illuminates this relationship, relying on the effective moment method, viz. Equation (2.6), for calculation of the torque of electrical origin.

Electrically induced particle rotation provides one explanation for the observed disruption that sometimes occurs during the initial chain formation step of electrofusion procedures (Holzapfel et al., 1982). If the linearly polarized AC electric field is within certain limited frequency ranges, linear chains of cells deform into irregular corkscrew configurations that interfere with electrofusion protocols. In some cases, the cells from disrupted chains spin steadily in the presence of the electric field. Zimmermann and Vienken (1982) hypothesized that this field-induced rotation, evident only when cells are proximate to one another, is responsible for the frequency-dependent disruption of chains. Though initially classified as a nuisance, *electrorotation* has become a very powerful diagnostic technique for measurement of certain critical dielectric properties of cells. Like dielectrophoretic levitation, electrorotation offers a means to study single cells in vitro, and it is sufficiently important to merit a detailed review in Section 4.6.

83

Certain differences of opinion have existed about the conditions required for electrorotation of cells and other small particles. In 1971, Pohl and Crane (1971) reported the observation of yeast cells rotating in a linearly polarized AC electric field. Pohl came to call this phenomenon *cellular spin resonance*. More recently, workers in the field of electrorotation questioned whether or not such rotation – in a linearly polarized electric field – is possible. Their skepticism stemmed from the recognition that the Born–Lertes effect is observed only when a *rotating* electric field is used; furthermore, no others besides Pohl ever reported rotation under similar conditions. But, in 1987, Turcu published an analysis revealing that steady (±) particle rotation about some axis perpendicular to an AC linearly polarized electric field is indeed possible under certain conditions. These predictions, though still unconfirmed experimentally, are entirely consistent with the roughly analogous case of a single-phase, capacitor-start induction motor. Given an initial push, the rotor of such a machine will rotate in either direction, following the right- or left-circularly polarized component of the net magnetic field. In fact, the steady-state operation of most induction motors with single-phase excitation is based on this principle. The questions about Pohl's claim will probably never be resolved, yet the possibility that he indeed was first to observe field-induced rotation of cells in a linearly polarized electric field cannot be ruled out. On the other hand, the term "cellular spin resonance" is misleading, as no resonance need be invoked to explain the phenomenon. It is made quite clear in the following analysis that the mechanism responsible for rotation is a relaxation process (Maxwell–Wagner polarization) that retards development of the dipole moment and creates a nonzero torque when the electric field rotates.

4.2 Theory for particle rotation

We can provide the necessary groundwork for an explanation of the Born–Lertes effect by deriving an expression for the electrical torque exerted on a particle by a rotating field. Generality is assured by using the effective dipole moment, which permits the formulation of torque expressions for spherical shells with multiple layers. We adapt this approach in Section 4.4 to model the special case of spontaneous rotation in a DC electric field (Quincke rotation).

A. *Particle interactions with a rotating electric field*

Consider a homogeneous sphere with radius R, permittivity ε_2, and conductivity σ_2 in a liquid of permittivity ε_1, conductivity σ_1, and dynamic viscosity η_1. The particle experiences a counterclockwise (CCW) rotating circularly polarized electric field $\bar{E}(t)$.

$$\bar{E}(t) = E_0(\hat{x} \cos \omega t + \hat{y} \sin \omega t) \qquad (4.1)$$

where $\omega > 0$ is the angular frequency in radians per second. Figure 4.1 shows a simple arrangement of electrodes that, when excited by polyphase AC voltages, creates such a field. For convenience in following the subsequent analysis, we rewrite below the vector phasor expression for $\bar{E}(t)$ from Section 2.4C.

$$\underline{\bar{E}} = E_0(\hat{x} - j\hat{y}) \qquad (2.49)$$

Employing the principle of superposition, the effective moment induced in the stationary particle by this rotating field is

$$\underline{\bar{p}}_{eff} = 4\pi\varepsilon_1 R^3 \underline{K}(\omega)E_0(\hat{x} - j\hat{y}) \qquad (4.2)$$

where $\underline{K}(\omega)$ is the complex Clausius–Mossotti function given by Equation (2.30).

Equation (4.2) means that the effective moment vector of the particle $\bar{p}_{eff}(t)$, rotating synchronously with the electric field vector $\bar{E}(t)$, lags behind by the constant angle α, as illustrated in Figure 4.2. The phase factor associated with $\underline{K}$ in Equation (4.2) is α, and it determines the magnitude and sign of the time-averaged torque.

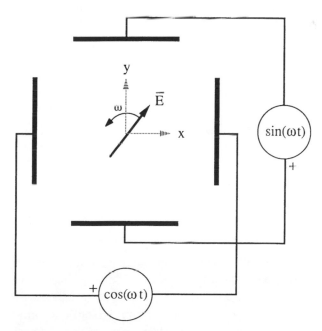

Fig. 4.1 Set of electrodes with polyphase AC excitation used to create a counterclockwise (CCW) rotating electric field.

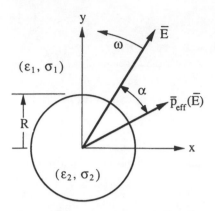

Fig. 4.2 Rotating electric field $\bar{E}$ and effective electric dipole moment $\bar{p}_{\text{eff}}$ vectors with lag angle α.

$$\langle \bar{T}^e(t) \rangle = \frac{1}{2}\text{Re}[\,\bar{p}_{\text{eff}} \times \underline{\bar{E}}^*] = |\bar{p}_{\text{eff}}| E_0 \sin \alpha \hat{z} \tag{4.3a}$$

$$\alpha = -\sin^{-1}\{\,\text{Im}[\underline{K}(\omega)]/|\underline{K}(\omega)|\,\} \tag{4.3b}$$

where $|\bar{p}_{\text{eff}}|$ is the constant magnitude of the rotating effective moment vector. Equations (4.3a,b) are also valid for any spherical shell, if the correct expression for the effective value of the particle permittivity ε_2' from Appendix C is used in $\underline{K}$.

B. Electric torque on homogeneous dielectric particle with ohmic loss

For a stationary homogeneous dielectric sphere with ohmic conductivity, we may employ either Equations (4.3a,b) or Equation (2.50) to derive the following expression for the torque:

$$\bar{T}^e = -\frac{6\pi\varepsilon_1 R^3 E_0^2(1 - \tau_1/\tau_2)\omega\tau_{\text{MW}}}{(1 + 2\varepsilon_1/\varepsilon_2)(1 + \sigma_1/2\sigma_2)[\,1 + (\omega\tau_{\text{MW}})^2]} \tag{4.4}$$

The sign of $\bar{T}^e$ depends on the relative values of the charge relaxation times of the particle τ_2 and the suspension medium τ_1. If $\tau_2 < \tau_1$, the particle rotates with the electric field, while if $\tau_2 > \tau_1$, the particle rotates in the direction opposite to the field. We can explain this sign dependence readily by examining the polarity of the sinusoidal distribution of *free* electric charge that accumulates on the surface of the particle. Refer to Figures 4.3a and b. The electrical charges on the

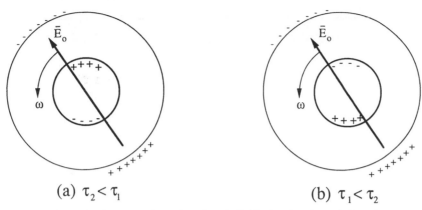

(a) $\tau_2 < \tau_1$ (b) $\tau_1 < \tau_2$

Fig. 4.3 Electrode charges and rotating electric field, showing the polarity of the induced free electric charge on a conductive spherical particle for two cases: (a) $\tau_2 < \tau_1$; (b) $\tau_2 > \tau_1$.

periphery represent the electrode polarity used to establish the imposed rotating field, while those just inside the particle are the induced charges upon which the rotating field acts to create the torque. Remember that the effective moment vector $\bar{p}_{eff}$ and the associated free charge always rotate synchronously with the electric field, irrespective of the relative magnitudes of τ_1 and τ_2. When $\tau_2 < \tau_1$, the sign of the induced charge in the particle is opposite to that in the electrodes, resulting in a positive torque. On the other hand, when $\tau_2 > \tau_1$, the induced charges in the particle are of the same sign as those in the electrodes. The continuous repulsion of like charge creates a time-average torque that leads to negative torque on the particle, even though the induced surface charge follows the field.

Now consider the more general case of a particle that is itself rotating at some angular velocity Ω. Assume that the particle is immersed in an ohmic dielectric liquid with dynamic viscosity η_1 and is subject to the rotating electric field of Equation (4.1). In the frame of reference of the particle, the effective electrical frequency is shifted as follows, $\omega \rightarrow (\omega - \Omega)$, and the torque expression must be altered accordingly.

$$\bar{T}^e = -\frac{6\pi\varepsilon_1 R^3 E_0^2(1 - \tau_1/\tau_2)(\omega - \Omega)\tau_{MW}}{(1 + 2\varepsilon_1/\varepsilon_2)(1 + \sigma_1/2\sigma_2)\left[1 + [(\omega - \Omega)\tau_{MW}]^2\right]} \qquad (4.5)$$

The steady-state equilibrium condition will be defined by a balance of this torque and the viscous retarding torque T^η.

$$T^\eta + T^e = 0 \qquad (4.6)$$

For the electrorotation rates typical of small particles with diameter less than $\sim500\,\mu\text{m}$, the laminar viscous torque formula may be invoked (Lamb, 1945, Sec. 334).

$$T^{\eta} = -8\pi\eta_1 R^3 \Omega \tag{4.7}$$

Solution of Equation (4.6) for the steady-state angular velocity Ω is represented by intersections of the electric torque–speed curve $T^e(\Omega)$ with the viscous load line, $-T^{\eta}(\Omega)$. Refer to Figures 4.4a and b, which plot representative curves for these two quantities. Note that as long as $\tau_1 > \tau_2$, there will always be a single solution to Equation (4.6), such that $\Omega > 0$. When $\tau_2 > \tau_1$, there will be a solution such that $\Omega < 0$; however, two more solutions $\Omega > \omega$ will exist if the viscosity of the suspending medium η_1 is not too high. The stability of any rotational equilibrium, that is, $\Omega = \Omega_{eq}$, is determined by the following set of conditions:

$$\text{If } \left.\frac{\partial(-T^{\eta})}{\partial\Omega}\right|_{\Omega_{eq}} > \left.\frac{\partial(T^e)}{\partial\Omega}\right|_{\Omega_{eq}}, \text{then } \Omega_{eq} \text{ is stable} \tag{4.8a}$$

$$\text{If } \left.\frac{\partial(-T^{\eta})}{\partial\Omega}\right|_{\Omega_{eq}} > \left.\frac{\partial(T^e)}{\partial\Omega}\right|_{\Omega_{eq}}, \text{then } \Omega_{eq} \text{ is unstable} \tag{4.8b}$$

Therefore, only one of the two equilibria for $\Omega > \omega$ shown in Figure 4.4b is stable.

The torque–speed curve of Figure 4.4a is virtually identical in form to that for a rotating induction machine. If $\tau_2 < \tau_1$ and if the viscosity η_1 is low (or, equivalently, if the electric field magnitude E_0 is high), then the particle rotates close to synchronism, that is, $\Omega \approx \omega$. As the viscous load is increased, Ω decreases. This behavior is very similar to magnetic induction interactions. On the other hand, if $\tau_2 > \tau_1$ as depicted in Figure 4.4b, the particle can rotate opposite to the circularly polarized electric field vector. There is no analogy to this counter-rotational behavior in a conventional magnetic induction machine with polyphase excitation.

C. Turcu's bifurcation theory

In 1987, Turcu published a mathematical analysis of the problem of a homogeneous dielectric sphere with finite conductivity in a uniform, linearly polarized AC electric field. Starting with the effective moment approach of Equation (2.48), he examined the stability of static and dynamic rotational equilibria of a homogeneous dielectric sphere and showed that, under certain conditions, a spherical particle can exhibit stable rotational equilibria in a linearly polarized AC electric field.

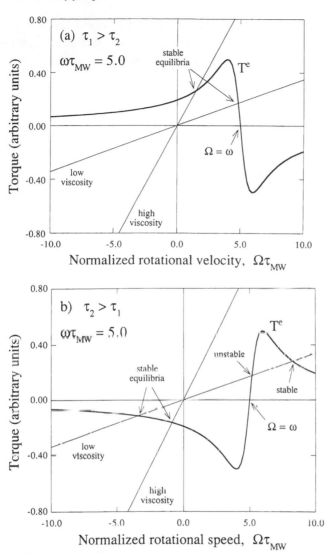

Fig. 4.4 Representative electrical torque–speed curve and viscous load line curves for homogeneous dielectric, ohmic spherical particle in a rotating electric field with normalized electrical frequency $\omega\tau_{MW} = 5.0$. Intersections represent stable or unstable rotational equilibria. The direction of rotation for the principal stable equilibria is determined by the relative magnitudes of the intrinsic charge relaxation times. (a) If $\tau_2 < \tau_1$, then $\Omega > 0$; (b) if $\tau_2 > \tau_1$, then $\Omega < 0$.

The starting point for Turcu's treatment is the deceptively simple problem of a homogeneous dielectric sphere with ohmic conductivity rotating at some angular velocity Ω in a linearly polarized AC electric field.

$$\bar{E}(t) = \hat{x}E_0 \cos \omega t \tag{4.9}$$

Using superposition, he decomposed the electric field into two circularly polarized vectors of equal magnitude, rotating in opposite directions.

$$\bar{\underline{E}}(t) = \frac{E_0}{2}(\hat{x} - j\hat{y}) \left[\exp (j\omega t) + \exp (-j\omega t) \right] \tag{4.10}$$

In the frame of reference rotating with the sphere at angular velocity Ω, the two rotating electric field components of Equation (4.10), designated by subscripts "+" and "−", become

$$\bar{\underline{E}}_{\pm}(t) = E_0(\hat{x} - j\hat{y}) \exp \left[j(\pm \omega - \Omega) t \right] \tag{4.11}$$

where $\bar{\underline{E}}(t) \equiv \frac{1}{2} (\bar{\underline{E}}_+ + \bar{\underline{E}}_-)$. In the frame of reference of the particle, the two electric field vectors appear to rotate at different frequencies. The two counter-rotating components of the induced effective moment vector are then

$$\bar{\underline{p}}_{\text{eff},\pm}(t) = 4\pi\varepsilon_1 R^3 \underline{K}_{\pm} E_0(\hat{x} - j\hat{y}) \exp \left[j(\pm \omega - \Omega)t \right] \tag{4.12}$$

where $\bar{\underline{p}}_{\text{eff}}(t) \equiv \frac{1}{2} (\bar{\underline{p}}_{\text{eff},+} + \bar{\underline{p}}_{\text{eff},-})$. It is necessary to define two complex Clausius–Mossotti functions.

$$\underline{K}_{\pm} \equiv \frac{\varepsilon_2^{\pm} - \varepsilon_1^{\pm}}{\varepsilon_2^{\pm} + 2\varepsilon_1^{\pm}} \tag{4.13}$$

where

$$\varepsilon_1^{\pm} \equiv \varepsilon_1 + \sigma_1/j (\pm \omega - \Omega) \quad \text{and} \quad \varepsilon_2^{\pm} \equiv \varepsilon_2 + \sigma_2/j(\pm \omega - \Omega) \tag{4.14}$$

Figure 4.5, plotting the various components of the electric field and the dipole moment in the rest frame, affords a simple physical interpretation of this situation. The two oppositely rotating electric field components are symmetric about the x axis; however, due to the particle's rotation, the two components of the dipole moment are symmetrical in neither magnitude nor phase.

In Turcu's nomenclature, the net instantaneous electrical torque exerted on the particle by the field may be calculated using the following equation:

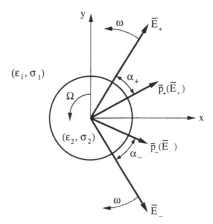

Fig. 4.5 Partition of linearly polarized AC electric field into CCW and CW circularly polarized rotating electric field vectors $\bar{E}_+$ and $\bar{E}_-$ with the respective moment vectors $\bar{p}_+$ and $\bar{p}_-$ induced by these components.

$$\bar{T}^e = \frac{1}{4}\mathrm{Re}[\bar{p}_{\mathrm{eff}}(t)] \times \mathrm{Re}[\bar{\underline{E}}(t)] \tag{4.15}$$

It should be evident that the cross terms in Equation (4.15) have zero time-average. Furthermore, any oscillatory contributions to the rotational dynamics may be justifiably ignored as long as $|\Omega| \ll \omega$. Under this assumption, which is ordinarily quite reasonable for particles smaller than $100\,\mu\mathrm{m}$ in diameter, the torque expression is considerably simplified.

$$T^e = 3\pi R^3 \varepsilon_1 F_0^2 \frac{(\tau_2 - \tau_1)\sigma_1\sigma_2}{(\varepsilon_2 + 2\varepsilon_1)(\sigma_2 + 2\sigma_1)}\left[\frac{(\Omega + \omega)\tau_{\mathrm{MW}}}{\tau_{\mathrm{MW}}^2(\Omega + \omega)^2 + 1} + \frac{(\Omega - \omega)\tau_{\mathrm{MW}}}{\tau_{\mathrm{MW}}^2(\Omega - \omega)^2 + 1}\right] \tag{4.16}$$

Turcu uses Equation (4.16) to obtain an expression for the equation of motion for the rotating sphere, then solves for the equilibria and determines their stability. Three distinct regimes can be identified:

1) A regime where only the static solution is allowed, that is, $\Omega = \Omega_1 = 0$;
2) A regime where only rotational solutions are allowed, that is, $\Omega = \pm\Omega_2$;[1] and
3) A regime where both static and rotational solutions are allowed, that is, $\Omega = 0, \pm\Omega_3^1$.

We may illuminate the physical significance of these regimes by examining the dynamic equilibrium condition of Equation (4.6), using Equation (4.16) for

[1] Explicit expressions for Ω_2 and Ω_3 may be found in Turcu (1987).

Table 4.1 *Regimes of stable rotational equilibria for homogeneous dielectric sphere of permittivity* ε_2 *and conductivity* σ_2 *in linearly polarized electric field of radian frequency* ω

	$\omega\tau_{MW} > 1$	$\omega\tau_{MW} < 1$
$\tau_2 > \tau_1$	High viscosity: $\Omega = \Omega_1 = 0$ Low viscosity: $\Omega = 0, \pm\Omega_3$	High viscosity: $\Omega = \Omega_1 = 0$ Intermediate viscosity: $\Omega = 0, \pm\Omega_3$ Low viscosity: $\Omega = \pm\Omega_2$
$\tau_2 < \tau_1$	High viscosity: $\Omega = \Omega_1 = 0$ Intermediate viscosity: $\Omega = 0, \pm\Omega_3$ Low viscosity: $\Omega = \pm\Omega_2$	All viscosity: $\Omega = \Omega_1 = 0$

$T^e(\Omega)$ and Equation (4.7) for T^η. We again distinguish between the two important cases: $\tau_2 > \tau_1$ and $\tau_2 < \tau_1$ and, furthermore, explore the influence of the normalized frequency $\omega\tau_{MW}$ on the allowed rotational modes. Figures 4.6a and b illustrate the four important regimes defined by τ_2/τ_1 and $\omega\tau_{MW}$. Intersections of $T^e(\Omega)$ with the linear viscous load line $-T^\eta(\Omega)$ represent the dynamic equilibria for a particle. The stability of these rotational equilibria can be determined by application of the criteria found in Equations (4.8a,b). By adjusting the suspension viscosity between low and high values, we can reveal the behavior of the equilibria.

Table 4.1, which summarizes the possible stable rotational equilibria obtained from examining Figures 4.6a and b, reflects all the important predictions of the original theory (Turcu, 1987).

4.3 Rotational (relaxation) spectra

The field-driven spinning of a particle in a rotating electric field can be exploited to reveal important information about its dielectric properties. The technique is called *electrorotation,* and its implementation is quite simple. The particle rotation speed Ω is measured as a function of the electrical frequency ω, usually with the electric field strength fixed. The resulting plot of Ω versus ω, called the rotational (ROT) spectrum, reveals important information about certain particle parameters. The application of this method in the study of biological particles such as cells and protoplasts is described in Section 4.6.

A. General theory

Consider a multilayered spherical particle of radius R in a rotating electric field. We now use Equation (2.34) for $\underline{K}(\omega)$ and, assuming that the electrical fre-

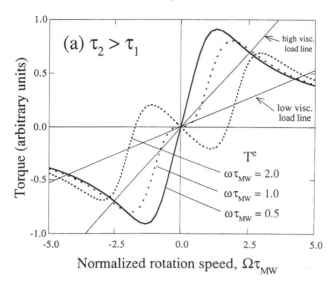

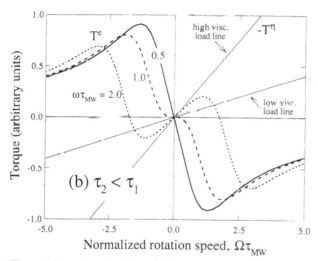

Fig. 4.6 Representative electrical torque–speed curve and viscous load line curves for homogeneous dielectric sphere with ohmic loss in a linearly polarized AC electric field for $\omega\tau_{MW} = 0.5, 1, 2$. (a) $\tau_2 > \tau_1$; (b) $\tau_2 < \tau_1$.

quency is much greater than the rotational speed, $\omega \gg |\Omega|$, calculate the time-average torque. First, we combine Equation (2.50) with Equations (4.6) and (4.7) to obtain a general expression for the rotational speed.

$$\Omega = -\frac{\varepsilon_1 \mathrm{Im}[\underline{K}(\omega)] E_0^2}{2\eta_1} \tag{4.17}$$

Equation (4.17) expresses the ROT spectrum: that is, particle rotation speed Ω as a function of the electric field frequency ω. For the simple though important case of a homogeneous dielectric sphere with ohmic loss, we have

$$\Omega = -\frac{3\varepsilon_1 \sigma_1 \sigma_2 (\tau_2 - \tau_1) E_0^2}{2\eta_1 (\sigma_2 + 2\sigma_1)(\varepsilon_2 + 2\varepsilon_1)} \left[\frac{\omega\tau_{MW}}{(\omega\tau_{MW})^2 + 1} \right] \qquad (4.18)$$

We note that the rotational spectrum for the homogeneous sphere has a single peak at the Maxwell–Wagner frequency: $\omega = 1/\tau_{MW}$. Just as predicted by Figures 4.4a and b, the direction of rotation depends upon the relative magnitudes of τ_1 and τ_2.

For a layered particle, $\mathrm{Im}[\underline{K}(\omega)]$ in Equation (4.17) will reflect a rich complexity of relaxations – one for each interface plus one for each intrinsic dielectric relaxation. When these relaxations are close together, individual peaks become difficult or impossible to distinguish and no convenient expression for $\Omega(\omega)$ may be formulated. On the other hand, if the relaxation time constants of the effective moment p_{eff} are well-separated (in other words, if $\tau_\alpha \gg \tau_\beta \gg \tau_\gamma \gg \ldots \gg \tau_N$), then we may use Equation (2.35b) to replace $\mathrm{Im}[\underline{K}(\omega)]$.[2]

$$\Omega(\omega) = -\frac{\varepsilon_1 E_0^2}{2\eta_1} \left[\frac{\omega\tau_\alpha \Delta K_\alpha}{\omega^2 \tau_\alpha^2 + 1} + \frac{\omega\tau_\beta \Delta K_\beta}{\omega^2 \tau_\beta^2 + 1} + \cdots + \frac{\omega\tau_N \Delta K_N}{\omega^2 \tau_N^2 + 1} \right] \qquad (4.19)$$

According to Equation (4.19), the spectrum of a multilayered particle consists of a set of positive and negative rotational peaks, each centered at a frequency equal to the reciprocal time constant of a relaxation, that is, $1/\tau_\alpha$, $1/\tau_\beta$, ..., $1/\tau_N$. Consistent with the Kramers–Krönig relations, Equations (E.10) and (E.11) in Appendix E, these rotational peaks are positive (+ROT) when $\Delta K_i < 0$ and negative (–ROT) when $\Delta K_i > 0$.

B. Sample spectra

The ROT spectra for homogeneous spheres shown in Figures 4.7a and b have been calculated using the same permittivity and conductivity values used to plot $\mathrm{Re}[\underline{K}(\omega)]$ in Figures 3.2a and b, respectively. In each case, only one rotational peak exists because there is only a single interface; the relaxation process is simply Maxwell–Wagner interfacial polarization. To facilitate comparison of DEP to ROT spectra, we have superimposed $\mathrm{Re}[\underline{K}(\omega)]$ in these figures. The permittivity and conductivity values in Figure 4.7a are such that the particle exhibits

---

[2] Refer to Appendix E for a general definition of the time constants τ_α, τ_β, τ_γ etc.

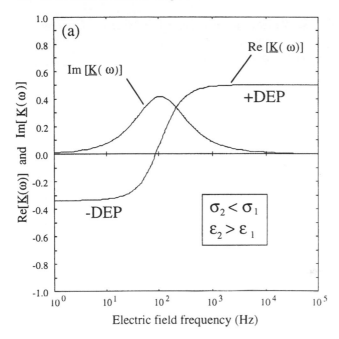

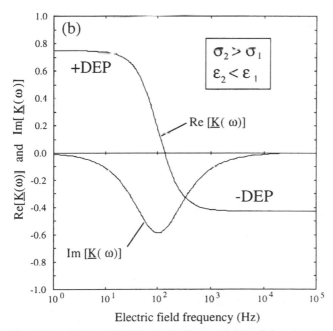

Fig. 4.7 Im[$\underline{K}$] (and Re[$\underline{K}$]) plot for homogeneous dielectric sphere with ohmic conductivity versus frequency: (a) $\varepsilon_2/\varepsilon_1 = 4$, $\sigma_2/\sigma_1 = 0.25$; (b) $\varepsilon_2/\varepsilon_1 = 0.1$, $\sigma_2/\sigma_1 = 10$.

negative dielectrophoresis at low frequencies and positive dielectrophoresis at high frequencies, whereas in Figure 4.7b positive DEP reigns at low frequencies and negative DEP at high frequencies. As expected, the rotational peaks are negative (counter-rotating) and positive (co-rotating), respectively, and occur at $\omega = (\tau_{MW})^{-1} = (\sigma_2 + 2\sigma_1)/(\varepsilon_2 + 2\varepsilon_1)$. At this frequency, $Re[\underline{K}(\omega)]$ equals the arithmetic mean of the high- and low-frequency limits, in other words, $Re[\underline{K}(\omega_{MW})] = (K_0 + K_\infty)/2$.

Figure 4.8 shows the calculated rotational spectra at several values of σ_1, the conductivity of the suspension medium, of a more complex model particle representing a walled yeast cell. The cell parameters are identical to those of Figure 3.3. Two relaxation peaks for the walled cells are evident, while a third is almost indistinguishable because the relaxation frequencies are not very well separated.

Figure 4.9 contains plots of $Im[\underline{K}]$ for a model protoplast cell, using the parameters defined and used in Figure 3.4. These spectra, again plotted for several values of suspension conductivity, reveal how the various critical frequencies depend upon σ_1. The strong dependence of the ROT peak upon suspension conductivity enhances the value of electrorotation in the investigation of cell

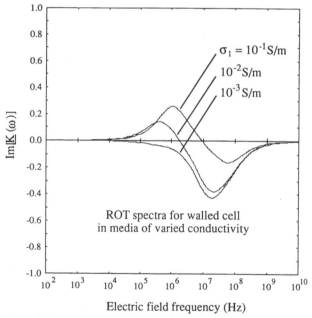

Fig. 4.8 Rotational spectra ($Im[\underline{K}]$) for model walled cell with lossless membrane in suspension media of varied conductivity. Parameters used are identical to those used in Figures 3.3 and E.5.

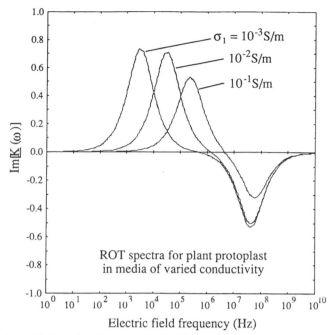

Fig. 4.9 Rotational spectra (Im[$\underline{K}$]) for model plant protoplast with lossless membrane in suspension media of varied conductivity. Parameters used are identical to those used in Figures 3.4 and E.4.

properties. The protoplast model is sufficiently important to warrant closer examination using Equation (E.4) for $\underline{K}(\omega)$. By the judicious use of some physically-based approximations, we obtain

$$\underline{K}(\omega) \approx K_{\infty} - \frac{\Delta K_\alpha}{j\omega\tau_\alpha + 1} - \frac{\Delta K_\beta}{j\omega\tau_\beta + 1} \tag{4.20}$$

where, if $\sigma_c \gg \sigma_1$ and $c_m R \gg \varepsilon_1$, we have

$$K_\infty = \frac{\varepsilon_c - \varepsilon_1}{\varepsilon_c + 2\varepsilon_1}, \quad \Delta K_\alpha \approx 1.5, \quad \Delta K_\beta \approx -\frac{3\varepsilon_1}{\varepsilon_c + 2\varepsilon_1} \tag{4.21a}$$

$$\tau_\alpha \approx \frac{c_m R}{2\sigma_1} \quad \text{and} \quad \tau_\beta \approx \frac{\varepsilon_c + 2\varepsilon_1}{\sigma_c} \tag{4.21b}$$

It should be clear now that we can extract estimates for membrane capacitance c_m and cytoplasmic conductivity σ_c from the electrorotational spectra of protoplasts, using the technique of varied suspension conductivity σ_1 to improve the precision. Some of these investigations are reviewed in Section 4.6A.

C. Argand diagrams

Because the dielectrophoretic force and rotational torque are directly proportional to $\text{Re}[\underline{K}(\omega)]$ and $\text{Im}[\underline{K}(\omega)]$, respectively, it is apparent that the DEP and ROT spectra are intimately related. Fuhr (1986) explored this relationship by plotting trajectories of $\underline{K}(\omega)$ in the complex plane and confirmed that these trajectories take the form of linked semicircles when the relaxation frequencies are well-separated. These complex maps – called Argand diagrams – are analogous to the well-known Cole–Cole plots often employed in the characterization of lossy dielectric media (Coelho, 1979). It is not surprising that the real and imaginary parts of $\underline{K}(\omega)$ should exhibit such a close interrelationship because, as a consequence of the linear dependence of $\bar{p}_{\text{eff}}$ upon $\bar{E}$, $\text{Re}[\underline{K}(\omega)]$ and $\text{Im}[\underline{K}(\omega)]$ must satisfy the Kramers–Krönig integral relations (Landau and Lifshitz, 1960). Argand diagrams of $\underline{K}(\omega)$ are valuable because they provide concise simultaneous graphical representation of the often complex dielectrophoretic and rotational behavior of multilayered particles. Refer to Appendix E, which summarizes the allowable forms that can be taken by $\underline{K}(\omega)$. This appendix also provides a set of simple rules for the Argand diagrams of layered spherical particles, making it easier to predict the DEP spectrum of a particle from its ROT spectrum and vice versa.

4.4 Quincke rotation in DC electric field

The Quincke effect is a spontaneous rotation exhibited by small particles suspended in liquid when subjected to a DC electric field exceeding some threshold value. The rotational axis is always perpendicular to the electrostatic field of force. After an initial startup transient, the rotation settles to a constant angular velocity proportional to the square of the electric field magnitude. As in the case of a rotating electric field, the relative magnitudes of the two charge relaxation times τ_1 and τ_2 are important; in fact, Quincke rotation is observed only when $\tau_2 > \tau_1$, an observation first made by Lampa (1906). Though seemingly quite distinct from particle rotation in an AC rotating field, the Quincke effect is in fact closely related and may be treated as a special case of electrorotation.

A. Theory

We can explore the threshold property of Quincke rotation by examining the stability of the static equilibrium (Melcher and Taylor, 1969; Jones, 1984). Imagine that the particle, initially at rest, is perturbed by a small rotational displacement. We wish to know if this perturbation will grow or be damped out by the viscosity. To proceed, assume that the particle is rotating at some small time-dependent angular velocity $\delta\Omega$ in a uniform DC electric field of magnitude E_0. In the frame of reference of the particle, the electric field appears to be rotating in the clockwise direction and, from Equation (4.5), the electrical torque is

$$T^e = \frac{6\pi\varepsilon_1 R^3 E_0^2 (1 - \tau_1/\tau_2)\delta\Omega\tau_{MW}}{(1 + 2\varepsilon_1/\varepsilon_2)(1 + \sigma_2/2\sigma_1)[1 + (\delta\Omega\tau_{MW})^2]} \tag{4.22}$$

The dynamic equation of motion is

$$I\frac{d(\delta\Omega)}{dt} = T^e(\delta\Omega) + T^\eta(\delta\Omega) \tag{4.23}$$

where $I = 8\pi\rho_2 R^5/15$ is the inertial moment of the spherical particle and ρ_2 is its mass density. Our concern is the stability of the solutions to Equation (4.23). If we assume that $|\delta\Omega|\tau_{MW} \ll 1$, then Equation (4.22) for the torque of electrical origin $T^e(\delta\Omega)$ may be linearized to obtain

$$I\frac{d(\delta\Omega)}{dt} = \left[\frac{6\pi\varepsilon_1 R^3 E_0^2(1 - \tau_1/\tau_2)\tau_{MW}}{(1 + 2\varepsilon_1/\varepsilon_2)(1 + \sigma_2/2\sigma_1)} - 8\pi\eta_1 R^3\right]\delta\Omega \tag{4.24}$$

The solution for $\delta\Omega$ will be a growing exponential if the term in the brackets is positive, which is possible only when $\tau_2 > \tau_1$. The instability condition is that the electric field must exceed a threshold value obtained by equating the term in brackets to zero.

$$(E_0)_{critical} = \sqrt{\frac{8\eta_1\sigma_1\sigma_2}{3\varepsilon_1(\tau_2 - \tau_1)}}(1 + \sigma_2/2\sigma_1) \tag{4.25}$$

Note that this critical value of electric field is defined only if $\tau_2 > \tau_1$. Spontaneous Quincke rotation is observed when the electric field exceeds this threshold, that is, when $E_0 > (E_0)_{critical}$.

A very simple physical interpretation for Quincke rotation emerges from an examination of the distribution of free electric charge on the surface of the particle when the DC electric field is applied. As shown in Figures 4.10a, and b, the sign of this charge depends upon the relative magnitudes of the two intrinsic charge relaxation times. If $\tau_2 > \tau_1$, the adjacent charges are of like sign so that a repulsion mechanism is activated; the result is that any small initial rotational displacement *in either direction* is amplified. On the other hand, when $\tau_2 < \tau_1$, then the sign of the charge at the surface is opposite that in the adjacent electrode and the perturbation is damped out. It should be clear from this line of reasoning that the axis of Quincke rotation is perpendicular to $\bar{E}_0$, but that the direction of this rotation is indifferent to the electric field vector.

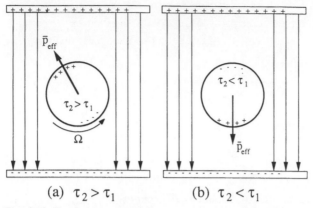

(a) $\tau_2 > \tau_1$ (b) $\tau_2 < \tau_1$

Fig. 4.10 Conductive dielectric particle in DC electric field showing the free electric charge induced at the surface of the particle. (a) When $\tau_2 > \tau_1$, the charge distribution is unstable and can lead to spontaneous Quincke rotation; (b) when $\tau_2 < \tau_1$, no spontaneous rotation occurs.

Steady rotation represents a balance achieved between the electrical and viscous torques.

$$T^e(\Omega_f) + T^\eta(\Omega_f) = 0 \qquad (4.26)$$

Plugging Equations (4.7) and (4.22) into this dynamic equilibrium condition gives an expression for the terminal angular velocity Ω_f in terms of E_0.

$$\Omega_f = \pm \tau_{MW}^{-1} \sqrt{\frac{E_0^2}{(E_0)_{critical}^2} - 1} \qquad (4.27)$$

The uncertainty of the sign of Ω_f reflects the fact that the direction of rotation depends entirely upon the initial disturbance.

B. Experimental results

Experimental data on Quincke rotation of small particles are sparse in the literature. Polystyrene spheres passively levitated in castor oil exhibit spontaneous rotation above a threshold value of the voltage when a DC electric field is used (Jones and Kallio, 1979). Unfortunately, in this experiment the levitation field is highly nonuniform, making quantitative study very difficult. Furthermore, finding suitable combinations of semi-insulating particles in the size range of 100 to 1000 μm and appropriate liquids to achieve simultaneously the conditions required for Quincke rotation, $\tau_2 > \tau_1$, and passive DC dielectrophoretic levitation, $\sigma_2 < \sigma_1$, is not easy.

4.5 Rotation of magnetizable particles

One can identify several distinct mechanisms leading to particle rotation in a magnetic field, including eddy current induction, hysteresis, and material anisotropy. Induction, the operative mechanism of most fractional horsepower AC electric motors, utilizes the eddy currents induced in a good (metallic) conductor of electricity when a time-varying or rotating magnetic field is present. Magnetic hysteresis, exploited in small synchronous motors, can create a net torque on a particle when the induced magnetic moment lags in time behind the applied rotating magnetic field. Anisotropy, a consequence of the crystalline structure of some magnetic materials, is still another torque-generation mechanism, though, as far as particle electromechanics is concerned, it probably has more influence on the orientation of particles and so is discussed in Chapter 5.

A. Induction

Consider a spherical particle with permeability μ_2, electrical conductivity σ_2, and radius R immersed in an electrically insulating fluid ($\sigma_1 = 0$) of permeability μ_1 and subjected to a CCW rotating magnetic field of frequency ω.

$$\vec{H}(t) = H_0(\hat{x} \cos \omega t + \hat{y} \sin \omega t) \tag{4.28}$$

or, in vector phasor form,

$$\underline{\vec{H}} = H_0(\hat{x} - j\hat{y}) \tag{4.29}$$

Starting with the expression for the effective magnetic dipole moment in a linearly polarized magnetic field from Section 3.5D, the rotating moment becomes

$$\underline{\vec{m}}_{\text{eff}} = 4\pi R^3 \underline{L} H_0(\hat{x} - j\hat{y}) \tag{4.30}$$

The frequency-dependent, complex coefficient $\underline{L}$ takes the place of the Clausius–Mossotti factor $\underline{K}$ used for dielectric particles. Figure 4.11 depicts the rotating magnetic field and effective moment vectors. Note that the magnetic moment always lags behind the rotating field by some angle α such that $0° < \alpha < 90°$, creating a positive torque, which causes rotation of the particle in the same direction as the field. Counter-rotation would be expected only in the case of a suspension medium with finite conductivity, that is, $\sigma_1 \neq 0$, where *magnetohydrodynamic flow* of the suspension medium would greatly complicate the rotational behavior.

Referring to Table A.2 from Appendix A, the general expression for the time-average torque is

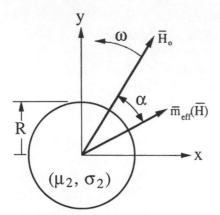

Fig. 4.11 Rotating magnetic field $\bar{H}$ and effective magnetic dipole moment $\bar{m}_{\mathrm{eff}}$ with lag angle α.

$$\langle \bar{T}^{m}(t) \rangle = \frac{\mu_1}{2} \mathrm{Re}[\,\bar{\underline{m}}_{\mathrm{eff}} \times \underline{\bar{H}}^*\,] \tag{4.31}$$

Using Equations (4.29) and (4.30) in Equation (4.31) yields the following expression for the torque due to eddy current induction on a spherical particle due to a rotating magnetic field.

$$\langle \bar{T}^{m}(t) \rangle = -4\pi\mu_1 R^3 \mathrm{Im}[\underline{L}] H_0^2 \hat{z} \tag{4.32}$$

Comparison of Equations (3.67) and (4.32) for the magnetophoretic force and the magnetic torque to Equations (2.45) and (2.48) for dielectrophoretic force and electrostatic torque reveals that the frequency-dependent factor $\underline{L}(\omega)$ plays a role similar to the Clausius–Mossotti factor $\underline{K}(\omega)$.

The frequency dependence of the time-average magnetic torque is contained within $\mathrm{Im}[\underline{L}(\omega)]$ and may be expressed in terms of the magnetic Reynolds number $R_{\mathrm{m}} = R\sqrt{\omega\mu_2\sigma_2/2}$ and the relative permeability $\mu_{\mathrm{R}} = \mu_2/\mu_1$.

$$\mathrm{Im}[\underline{L}] = \frac{3R_{\mathrm{m}}^2(C_- - R_{\mathrm{m}}S_+)}{(1-\mu_{\mathrm{R}})^2 R_{\mathrm{m}}^2 C_+ + [(1-\mu_{\mathrm{R}})^2/2 + 2\mu_{\mathrm{R}}^2 R_{\mathrm{m}}^4]C_- - (1-\mu_{\mathrm{R}})^2 R_{\mathrm{m}} S_+ + 2\mu_{\mathrm{R}}(1-\mu_{\mathrm{R}})R_{\mathrm{m}}^3 S_-} \tag{4.33}$$

where

$$C_{\pm}(R_{\mathrm{m}}) = \cosh(2R_{\mathrm{m}}) \pm \cos(2R_{\mathrm{m}}) \quad \text{and} \quad S_{\pm}(R_{\mathrm{m}}) = \sinh(2R_{\mathrm{m}}) \pm \sin(2R_{\mathrm{m}}) \tag{4.34}$$

Figure 4.12 shows the frequency-dependent behavior of Im[L] and, for reference, Re[L] as a function of R_m. Note how the low- and high-frequency limits conform to our expectations. Im[L] always exhibits a negative peak, indicating that the particle always rotates in the same direction as the magnetic field. Richards et al. (1981) have reported the rotation of magnetically levitated objects.

In the important case of a nonmagnetic conducting sphere ($\mu_2 = \mu_0$) in a nonmagnetic medium ($\mu_1 = \mu_0$), where only the diamagnetic effect of the eddy currents is present, Im[L] reduces to a more compact form.[3]

$$Im[L] = \frac{3}{4R_m^2} - \frac{3}{8R_m} \frac{\sinh 2R_m + \sin 2R_m}{\sinh^2 R_m + \sin^2 R_m} \tag{4.35}$$

B. Nonlinear effects

The behavior of a magnetizable but nonconducting particle in a rotating magnetic field depends upon whether the particle is magnetically hard or soft. Magnetically soft materials, as represented by the *M–H* curve of Figure 3.9a, have only modest hysteresis. As a result, the effective magnetic dipole moment remains essentially parallel to the rotating field and the magnetic torque will be very weak. On the other hand, a particle made of magnetically hard material will

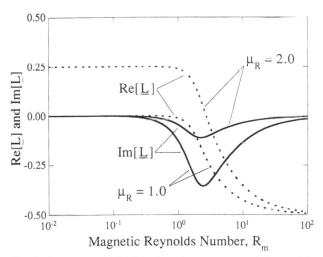

Fig. 4.12 Im[L] (and Re[L]) plotted versus magnetic Reynolds number R_m for $\mu_R = \mu_2/\mu_1 = 1$ and 2.

---

[3] Equation (4.35) is the companion of Equation (3.71) for the MAP force on a nonmagnetic conductive sphere.

exhibit behavior similar to either a hysteresis motor (assuming that the rotating field is sufficiently strong to induce saturation in the particle) or a permanent magnet motor.

Consider a spherical particle of magnetically hard material, as shown in Figure 3.9b, suspended in a fluid with linear permeability μ_1. Consistent with Appendix A.2 of Fröhlich (1958), the effective magnetic moment of the particle will be given by

$$\overline{m}_{eff} = 4\pi R^3 \frac{\mu_1}{\mu_0 + 2\mu_1} \overline{M}_{rem} + 4\pi R^3 \frac{\mu_0 - \mu_1}{\mu_0 + 2\mu_1} \overline{H}_0 \qquad (4.36)$$

If remanent magnetization predominates, then the magnitude of the vector moment $|\overline{m}_{eff}|$ will be fixed.

$$|\overline{m}_{eff}| \approx m_{rem} \equiv 4\pi R^3 \frac{\mu_1}{\mu_0 + 2\mu_1} M_{rem} \qquad (4.37)$$

This moment vector will rotate synchronously at fixed lagging angle α behind the rotating field while the particle smoothly gains speed and eventually locks into synchronism with the field. Analogous to a hysteresis motor, the torque is essentially constant from $\Omega = 0$ up to the synchronous speed $\Omega = \omega$, as shown in Figure 4.13. For such a particle rotating at synchronous speed in a liquid of viscosity η_1, the lag angle will be

$$\alpha = \sin^{-1}\left[\frac{8\pi\eta_1 R^3 \omega}{\mu_1 m_{rem} H_0}\right] \qquad (4.38)$$

The maximum magnetic torque, $T_{max} = \mu_1 m_{rem} H_0$, achieved at $\alpha = 90°$, limits the rotational frequency of the driving magnetic field.

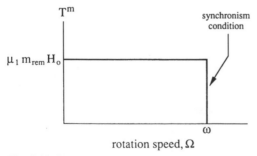

Fig. 4.13 Representative torque-speed curve for hysteretic magnetic particle in a rotating field.

$$\omega \le \omega_{max} = \frac{\mu_1 m_{rem} H_0}{8\pi\eta_1 R^3} \qquad (4.39)$$

At any magnetic field frequency above ω_{max}, the torque simply cannot sustain rotation and the particle stalls.

A magnetically hard particle with a permanent moment can rotate only at the synchronous speed, that is, $\Omega = \omega$, because $\bar{m}_{eff}$ is fixed to the particle. The starting torque is zero and, in order to achieve synchronism, there must be sufficient initial rotation to allow the particle to catch up to the synchronous condition. Similar to the case of hysteresis, the maximum torque for a permanently magnetized particle is limited by the value of the remanent moment m_{rem}. Equations (4.38) and (4.39) may be used to determine the lag angle α as well as a maximum frequency ω_{max}, above which stalling occurs.

4.6 Applications of electrorotation

Initial impetus for the revival of interest since 1980 in electric field–induced particle rotation was provided not by any proposed application, but rather by the observation that ordinarily stable chains of biological cells were disrupted when the AC electric field was adjusted within certain, fairly narrow frequency bands. This chain disruption was a significant nuisance in cell electrofusion procedures (Holzapfel et al., 1982). An investigation of the phenomenon revealed that the mutual interactions of two closely spaced cells or other particles could create a rotating field component that led to an electrical torque on each particle. The principal manifestations of the effect for such closely spaced particles are illustrated in Figure 4.14. Two particles in either parallel (‖) or perpendicular (⊥) alignment to the linearly polarized electric field exhibit no cooperative rotation; however, in any other alignment, both particles are observed to rotate if the electric field frequency is within certain narrow bands.[4] To understand this phenomenon, note first that the net electric field experienced by each particle consists of the imposed field plus the induced dipole field due to its neighbor. Maxwell–Wagner polarization delays the phase of the effective moment induced in both particles. Therefore, the first particle responds to the vector sum of the imposed field and the induced field from its neighbor, which is neither parallel to nor in phase with the imposed field. As a result, when two closely spaced particles are misaligned with respect to the electric field, each of the particles experiences an

---

[4] The cell rotation hypothesis is one of two possible explanations for frequency–dependent chain disruption. In Section 6.4, an alternate explanation for chain disruption, based upon frequency-dependent orientational instability, is advanced.

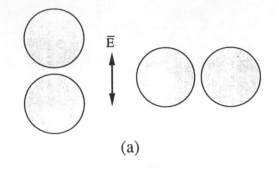

(a)

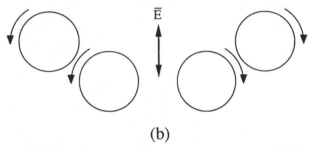

(b)

Fig. 4.14 Rotational interactions of two identical closely spaced dielectric particles with loss situated in a linearly polarized AC electric field.

(a) Particles in parallel or perpendicular alignment do not rotate, though they will attract or repel one another, respectively.

(b) Misaligned particles tend to rotate in the directions shown due to mutual interactions.

electric field with a rotating component. This rotating component will always be strongest near the peaks of $\text{Im}[\underline{K}(\omega)]$. Thus, each particle serves the function of *pole shading* for the other, creating the necessary condition for rotation. The problem of chain disruption in electrofusion protocols is solved by avoiding certain critical frequency ranges.

A. Cell characterization studies

The encounter with electrorotation as a nuisance in electrofusion has produced one very important positive outcome. Electrorotation is now the principal means used for the dielectric characterization of individual biological cells. Rotational spectra – that is, rotation speed versus electrical frequency at fixed voltage – can be analyzed to estimate fundamental cell properties such as membrane capacitance and conductance, and to monitor changes in these properties when cells are subjected to various treatments, protocols, or environmental insults. Both "before and after" measurements and continuous monitoring of individual cells are possible. Basic experimental electrorotation chambers are of simple construction, often consisting of a set of miniature electrodes arranged on a microscope slide to form a chamber with a fairly uniform electric field. Refer to Figure 4.1. The electrodes are driven by polyphase AC voltage waveforms (or properly phased square wave signals) to create a uniform rotating electric field at the center of the chamber. Individual cells are placed at the center of the chamber, with due care taken to avoid introducing any extra cells that would distort the rotating field. The cell's rotational velocity $\Omega/2\pi$ is measured as a function of the electric field frequency $f = \omega/2\pi$, ordinarily with the electric field strength held constant. It is usual to conduct additional measurements to test the expected square law dependence of the rotation speed Ω upon the electric field strength E_0. For typical cells on the order of 5 µm in diameter, electric field strengths of ~20 V/cm, and frequencies in the range from ~10^2 to ~10^8 Hz, the rotation speeds are of the order of a few revolutions per second.

Even when the relaxation frequencies for a given particle are well-separated, the broadness of electrorotational peaks makes it difficult to discriminate the electrical frequency where the maximum rotation speed occurs. A better approach is to superimpose two counter-rotating fields of different frequencies and precisely equal magnitudes. This method, due to Arnold (1988), leads to greatly improved measurement precision because broad rotational peaks are replaced by far more easily detected nulls. The necessary expression for Ω in terms of the two frequencies is readily obtained by a conventional superposition of the two torque terms into either Equation (4.18) or Equation (4.19).

Electrorotation offers improved resolution over standard low-frequency capacitance bridge techniques used on cell suspensions. The measurements are made on single cells and the technique does not depend on the use of approximate mixture formulas. On the other hand, electrorotation shares with DEP levitation the problem that, given the suspension medium conductivities required to maintain some types of cells, joule heating can be significant and disruptive fluid convection can occur. Thus, ionic concentrations of cell suspension media may rarely exceed ~5 mM. Also, electrode polarization problems can become important at frequencies less than approximately 1 kHz.

Investigators have used electrorotation to measure membrane properties such as capacitance per unit area c_m for a large variety of cells, including plant protoplasts (Arnold and Zimmermann, 1982), plant cell vacuoles and erythrocytes (Glaser et al., 1983), chloroplasts (Arnold et al., 1985), and oocytes (Arnold et al., 1989). More recently, Freitag and colleagues (1989) employed electrorotation to investigate the effects of centrifugation, hypertonic and hypotonic stress, and trypsin treatment on membrane structure. Fuhr et al. (1989) studied the phenomenon of reversible and irreversible membrane damage caused by varying the strength of the rotating electric field. Gimsa et al. (1991) reported measurements of the transient behavior of plant protoplasts in response to changes in suspension osmolarity.

Evidence of low-frequency dielectric anomalies emerges in the electrorotational spectra of latex spheres (Arnold et al., 1987). These particles, investigated because they mimic the surface charging behavior of biological cells, reveal positive electrorotation at low frequencies when none is expected (Kaler et al., 1992; Wang et al., 1992). Electrohydrodynamic convection in the counter-ion cloud may play some role in this anomalous behavior; such fluid mechanical interactions have not been considered in analyses published to date.

Washizu and his co-workers (1991) found a unique application of electrorotation in their investigations of the torque–speed behavior of the flaggellar motor mechanism of bacterial cells. They tethered oblong cells to a substrate by single flaggella, subjected them to a rotating electric field at ~500 kHz, and then observed their motion with a microscope. The same group reported a micromotor device achieving simultaneous levitation and electrorotation of small particles with a set of two six-pole planar electrodes and six-phase excitation (Washizu and Nakada, 1991).

B. Cell separation

Another innovative scheme, depicted in Figure 4.15, achieves separation of particles immersed in aqueous suspension and resting on an Agar surface by inducing the particles to roll in response to a rotating electric field (Fuhr et al., 1985). Depending upon the sign of $\text{Im}[\underline{K}(\omega)]$, cells will roll one way or the other. Thus, if the electric field frequency is controlled precisely, cells can be differentiated based on such factors as membrane capacitance c_m. A simple demonstration with plant protoplasts showed the feasibility of the concept.

C. Practical implications of the Quincke effect

Electrostatic repulsion motors operating on the principle of the Quincke effect have been built and tested (Sumoto, 1955, 1956; Secker and Scialom, 1968; Simpson and Taylor, 1971). These motors, of very simple design and construction, are high-voltage, low-current devices capable of delivering only modest mechanical power, usually less than ~10 W. Interest in Quincke rotation is likely to increase with the emphasis now being placed upon micromechanical transducers such as miniature pumps. Another possible application is in the processing

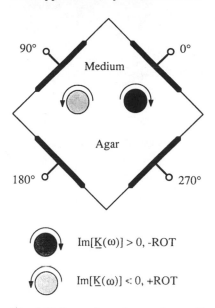

Im[$\underline{K}(\omega)$] > 0, -ROT

Im[$\underline{K}(\omega)$] < 0, +ROT

Fig. 4.15 Separation scheme where cells resting on an Agar surface rotate in response to the rotating electric field created by polyphase excitation of the electrodes (Fuhr et al., 1985).

and coating of small particles performed in semi-insulative liquids or plasmas. Here, a controllable rotation might be exploited in order to improve the uniformity of surface treatments applied to particles. Finally, Fuhr and Hagedorn have considered the Quincke effect as a possible energy transduction mechanism for the flagellae in motile bacteria (Fuhr and Hagedorn, 1987).

An interesting implication of the Quincke effect is that insulating contaminant particles, once they start to rotate, can increase the apparent bulk DC conductivity in dielectric liquids. This is because the rotation enhances charge transport around the particle. Presumably, this transport mechanism would be characterized by the field threshold described in Section 4.4A.

5

Orientation of nonspherical particles

5.1 Introduction

Elongated magnetizable particles respond to a magnetic field by aligning with their longest axes parallel to the field vector. This phenomenon is exploited in the familiar physics demonstration where iron filings are sprinkled on a sheet of paper placed atop a permanent magnet. As they land on the paper, the particles align themselves parallel to the field of the magnet and, if care is taken to distribute the particles uniformly, their pattern reveals the form of the magnetic field. Very similar behavior is manifested by elongated dielectric or electrically conductive particles in an electric field.

The physical origin of the *orientational torque* exerted on a particle is closely related to that of *rotational torque* discussed in Chapter 4 and, in fact, a particle may exhibit a combination of rotational dynamics and orientational behavior under certain conditions. There are, however, some important distinctions to be drawn between the two effects.

- An orientational torque acts only upon a particle exhibiting some type of *anisotropy* due to either intrinsic properties such as crystalline anisotropy or extrinsic factors such as shape. Rotational torque, on the other hand, can be experienced by a spherical *isotropic* particle.
- In a rotating field, the orientational torque can produce steady motion, but only at synchronous speed. In contrast, rotational torque induces steady rotation at sub-synchronous speed.
- Orientational torque does not depend on either dielectric or ohmic loss, while rotational torque requires it.
- Particles can experience orientational torque in both AC and DC electric fields. With the exception of the Quincke effect (c.f., Section 4.4), steady rotation is ordinarily observed only with an AC field.

The initial focus of this chapter is dielectric particles with ohmic conductivity having shape-dependent anisotropy and subjected to a uniform, linearly polarized

110

AC electric field. A close examination of this problem yields the important result that dielectric particle orientation is frequency-dependent and that, somewhat surprisingly, lossy particles do not always align with their longest axes parallel to the field. Frequency-dependent orientation is important because measurement of the *orientational spectra* of elongated cells such as certain erythrocytes and bacteria provides a means to study the dielectric properties of these cells (Miller and Jones, 1993).

Besides the orientational effects treated in this chapter, there is also the well-known tendency of an electric field to deform droplets and bubbles into ellipsoidal shape.[1] This phenomenon is the basis of the *bubble theory* of electrical breakdown (Garton and Krasucki, 1964). According to this theory, impurity gas bubbles (or liquid droplets) help precipitate breakdown in liquids because they create local regions of intensification of the electric field. Field-induced elongation exacerbates this field intensification. Electric field–induced droplet elongation may also be important as a mechanism in generating large numbers of submicron-sized, charged water droplets thought to be important in the dynamics of thunderstorm electrification. Certain biological cells also distort in the presence of an electric field (Friend et al., 1975; Engelhardt et al., 1984). The topic of bubble and droplet elongation is more appropriate for a treatise on continuum electromechanics and, while not irrelevant to the electromechanics of particles, is not considered in this book.

5.2   Orientation for lossless homogeneous ellipsoids

In this section, we use the effective moment method to investigate orientational phenomena and show that a lossless, isotropic, homogeneous, dielectric ellipsoid always responds to an applied electric field in the same way – by aligning itself with its longest axis parallel to the vector field. In Section 5.3, we generalize the analysis to account for finite particle and medium conductivity and discover that the orientation of a lossy ellipsoidal particle is a function of frequency.

A.   *Isotropic ellipsoid in a uniform electric field*

Consider a homogeneous ellipsoid having isotropic dielectric permittivity ε_2 and semi-axes a, b, and c to be aligned with the x, y, and z axes, respectively, as depicted in Figure 5.1. The particle is immersed in a dielectric fluid of

---

[1] Bubbles and droplets usually distort into prolate spheroidal shape with their long axes parallel to the electric field; however, oblate shapes can result for liquid droplets when electrohydrodynamic (EHD) motions are induced by the electric field (Allan and Mason, 1962; Torza et al., 1971).

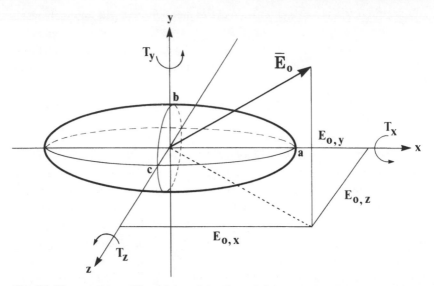

Fig. 5.1 Homogeneous ellipsoidal particle of permittivity ε_2 and ohmic conductivity σ_2 with semi-axes $a > b > c$ situated in a continuous medium having permittivity ε_1 and conductivity σ_1 and subjected to the uniform electric field $\bar{E}_0$. The indicated torque signs are consistent with the right-hand convention.

permittivity ε_1. For the sake of generality, we assume that the imposed uniform electric field $\bar{E}_0$ is not aligned with any of the three axes of the ellipsoid.

$$\bar{E}_0 = \hat{x}E_{0,x} + \hat{y}E_{0,y} + \hat{z}E_{0,z} \tag{5.1}$$

The objective is to find expressions for the three orthogonal components of the electrical torque using Equation (2.6). We therefore require an expression for $\bar{p}_{\text{eff}}$, the effective dipole moment of the particle. We can obtain this quantity by solving the electrostatic boundary value problem in ellipsoidal coordinates for the induced potential term and then comparing it to the general expression for the potential due to an equivalent dipole source. The procedure, based on a published solution of Stratton (Stratton, 1941, Sec. 3.29), is summarized in Appendix G.

An alternate method for identification of $\bar{p}_{\text{eff}}$ is predicated upon the fact that $\bar{E}^-$ the electric field inside the ellipsoid, though in general not parallel to $\bar{E}_0$, is uniform. According to this method, we express the effective moment in terms of the effective *excess polarization*.

$$\bar{p}_{\text{eff}} = \frac{4\pi abc}{3}(\varepsilon_2 - \varepsilon_1)\bar{E}^- \tag{5.2}$$

The components of the internal electric field $\bar{E}^-$ are related, term by term, to the imposed field $\bar{E}_0$ through a set of depolarization factors. For example, the x component is

$$E_x^- = \frac{E_{0,x}}{1 + \left(\dfrac{\varepsilon_2 - \varepsilon_1}{\varepsilon_1}\right)L_x} \tag{5.3}$$

with L_x defined by an elliptical integral.

$$L_x = \frac{abc}{2}\int_0^\infty \frac{ds}{(s + a^2)\,R_s} \tag{5.4}$$

where $R_s \equiv \sqrt{(s + a^2)\,(s + b^2)\,(s + c^2)}$. Similar expressions may be obtained relating E_y^- to $E_{0,y}$ and E_z^- to $E_{0,z}$ by appropriate substitutions of y (or z) for x and b (or c) for a, respectively. The observation that, in general, the imposed electric field ($\bar{E}_0$) and the uniform field inside the ellipsoid ($\bar{E}^-$) are not parallel is significant because the orientational torque arises only when $\bar{E}_0$ is not parallel to $\bar{p}_{\text{eff}}$. One can confirm that Equation (5.2) is identical to Equation (G.5) in Appendix G.

The depolarization factors L_x, L_y, and L_z are all positive and interrelated as follows:

$$0 \le L_\alpha \le 1, \text{ where } \alpha = x, y, \text{ or } z \tag{5.5a}$$

$$L_x + L_y + L_z = 1 \tag{5.5b}$$

In addition,

$$1 + \left(\frac{\varepsilon_2 - \varepsilon_1}{\varepsilon_1}\right)L_\alpha > 0, \text{ for } \alpha = x, y, \text{ or } z \tag{5.6}$$

With expressions for the components of the internal field, $\bar{E}^-$, we can evaluate the effective dipole moment vector using Equation (5.2), and then write expressions for the torque components. Knowledge of the signs of these torque components allows us to investigate the field-induced alignment behavior of ellipsoids and to arrive at some fairly general conclusions about the electric field–coupled orientation of any nonspherical particle.

B. Alignment torque expressions

To investigate the alignment behavior of dielectric ellipsoids in a uniform electrostatic field, we need to determine the signs of the three torque components:

T_x^e, T_y^e, and T_z^e. Combining Equation (2.6) with Equation (5.2), we obtain for the x component of torque,

$$
\begin{aligned}
T_x^e &= (p_{\text{eff}})_y E_{0,z} - (p_{\text{eff}})_z E_{0,y} \\
&= \frac{4\pi abc}{3}(\varepsilon_2 - \varepsilon_1)\,[E_y^- E_{0,z} - E_z^- E_{0,y}]
\end{aligned}
\tag{5.7}
$$

It is convenient here to replace Equation (5.7) by an expression that more clearly shows the dependence of T_x^e upon the relative magnitudes of certain important parameters. Substituting in for E_y^- and E_z^-, and then performing some algebraic manipulation, we obtain

$$
T_x^e = \frac{4\pi abc(\varepsilon_2 - \varepsilon_1)^2 (L_z - L_y) E_{0,y} E_{0,z}}{3\varepsilon_1\left[1 + \left(\dfrac{\varepsilon_2 - \varepsilon_1}{\varepsilon_1}\right)L_y\right]\left[1 + \left(\dfrac{\varepsilon_2 - \varepsilon_1}{\varepsilon_1}\right)L_z\right]}
\tag{5.8}
$$

with similar expressions for T_y^e and T_z^e. These same torque expressions result using an energy method (Stratton, 1941). The relative magnitudes of the depolarizing factors L_x, L_y, and L_z determine the signs of the three torque components, that is,

$$
T_x^e \propto (L_z - L_y)\,E_{0,y} E_{0,z}
\tag{5.9a}
$$

$$
T_y^e \propto (L_x - L_z)\,E_{0,z} E_{0,x}
\tag{5.9b}
$$

$$
T_z^e \propto (L_y - L_x)\,E_{0,x} E_{0,y}
\tag{5.9c}
$$

With no loss of generality, we now assume that the three imposed field components $E_{0,x}$, $E_{0,y}$, and $E_{0,z}$ are all positive. If we further assume that $a > b > c > 0$, then it is true that $0 < L_x < L_y < L_z < 1$ and, finally: $T_x^e > 0$, $T_y^e < 0$ and $T_z^e > 0$. Figure 5.1 depicts the sign conventions of these torque quantities. An examination of Equations (5.9a,b,c) reveals that the particle will achieve equilibrium when any of its three principal axes is aligned with $\bar{E}_0$; however, only one of these orientations is stable, that with the "a" axis (the longest) parallel to the electric field. Therefore, a freely suspended dielectric ellipsoid always tends to align itself with its longest axis parallel to the imposed field. This alignment behavior is completely independent of the relative magnitudes of ε_1 and ε_2; a particle will align with its longest axis parallel to $\bar{E}_0$ whether the DEP response is positive or negative.

C. Two special cases: prolate and oblate spheroids

Prolate and oblate spheroids merit attention because such shapes are common in technological applications where particle orientation is critical to the fabrication of composites with enhanced mechanical strength, directional dielectric properties, etc. The prolate spheroid ($a > b = c$) models cigar- or needle-shaped particles, while an oblate shape ($a = b > c$) represents lamellae and lenticular particles. Assume that the imposed electric field has components parallel (∥) and perpendicular (⊥) to the "a" axis of the particle, that is,

$$\bar{E}_0 = \bar{E}_{\parallel} + \bar{E}_{\perp} \tag{5.10}$$

Refer to Figures 5.2a and b, which depict prolate and oblate spheroids, respectively, with axis "a" at some angle θ with respect to $\bar{E}_0$.

The alignment torque may be calculated using the expression

$$\bar{T}^e = \bar{p}_{\parallel} \times \bar{E}_{\perp} + \bar{p}_{\perp} \times \bar{E}_{\parallel} \tag{5.11}$$

Using Equation (5.2) to express the components of the effective moment, the magnitude of the torque is

$$T^e = \frac{4\pi abc}{3}(\varepsilon_2 - \varepsilon_1)\,[E^-_{\parallel}E_{\perp} - E^-_{\perp}E_{\parallel}] \tag{5.12}$$

where

$$E^-_{\parallel} = \frac{E_{\parallel}}{1 + \left(\dfrac{\varepsilon_2 - \varepsilon_1}{\varepsilon_1}\right)L_{\parallel}} \quad \text{and} \quad E^-_{\perp} = \frac{E_{\perp}}{1 + \left(\dfrac{\varepsilon_2 - \varepsilon_1}{\varepsilon_1}\right)L_{\perp}} \tag{5.13}$$

The resulting equation for the net alignment torque looks very similar to Equation (5.8) but now is expressed in terms of the parallel and perpendicular components of the field.

$$T^e = \frac{4\pi abc\,(\varepsilon_2 - \varepsilon_1)^2\,(L_{\perp} - L_{\parallel})\,E_{\parallel}E_{\perp}}{3\varepsilon_1\left[1 + \left(\dfrac{\varepsilon_2 - \varepsilon_1}{\varepsilon_1}\right)L_{\parallel}\right]\left[1 + \left(\dfrac{\varepsilon_2 - \varepsilon_1}{\varepsilon_1}\right)L_{\perp}\right]} \tag{5.14}$$

C.1 Prolate spheroid

For prolate spheroids, $a > b = c$, so that $L_{\perp} = (1 - L_{\parallel})/2$. In this case, the elliptic integral used to evaluate $L_{\parallel}$ has an analytical solution (Stratton, 1941, Sec. 3.28).

$$L_{\parallel} = \frac{ab^2}{2}\int_0^{\infty} \frac{ds}{(s + a^2)^{3/2}(s + b^2)} = \frac{b^2}{2a^2 e^3}\left[\ln\left(\frac{1 + e}{1 - e}\right) - 2e\right] \tag{5.15}$$

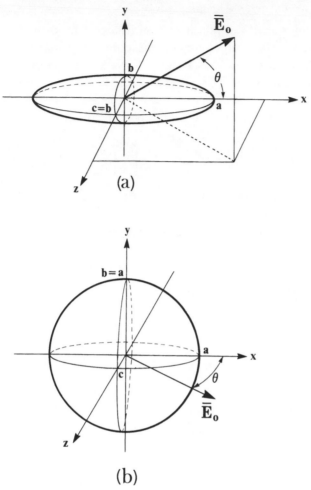

Fig. 5.2 Prolate (a) and oblate (b) spheroids of permittivity ε_2 and ohmic conductivity σ_2 with semi-axes a and b situated in a medium of permittivity ε_1 and conductivity σ_1 and subjected to the uniform imposed electric field $\bar{E}_0$. θ is the angle between semi-axis "a" and the imposed field $\bar{E}_0$.

where $e \equiv \sqrt{1 - b^2/a^2}$ is the eccentricity of the spheroid.[2]

For a highly elongated (needle-shaped) particle, $a \gg b = c > 0$, so that $L_{\parallel} \ll 1$ and $L_{\perp} \approx 1/2$. In this case, the torque expression simplifies considerably.

$$T^e_{\text{prolate}} \approx \frac{4\pi ab^2}{3} \varepsilon_1 E_{\parallel} E_{\perp} \frac{(\varepsilon_2 - \varepsilon_1)^2}{[\varepsilon_1 + (\varepsilon_2 - \varepsilon_1) L_{\parallel}] (\varepsilon_2 + \varepsilon_1)} \tag{5.16}$$

[2] A sign error found in one of the terms of Eq. (45), Sec. 3.28 of Stratton (1941) has been corrected in Equation (5.15).

For a needle-shaped particle, $L_{\parallel}$ may be approximated.

$$L_{\parallel} \approx \frac{b^2}{a^2}\left[\ln\left(\frac{2a}{b}\right) - 1\right] \tag{5.17}$$

By retaining the term involving $L_{\parallel} \ll 1$ in Equation (5.16), we can use the $\varepsilon_2/\varepsilon_1 \to \infty$ limit to derive an expression for the torque on a perfectly conducting prolate spheroid.

$$T^e \approx \frac{4\pi a^3}{3}\left[\frac{\varepsilon_1}{\ln(2a/b) - 1}\right]E_{\parallel}E_{\perp} \tag{5.18}$$

Using $E_{\parallel} = E_0\cos\theta$ and $E_{\perp} = E_0\sin\theta$, we obtain

$$T^e \approx \frac{2\pi a^3}{3}\left[\frac{\varepsilon_1}{\ln(2a/b) - 1}\right]E_0^2\sin 2\theta \tag{5.19}$$

which is identical to the expression obtained by Talbott and Stefanakos (1972).

C.2   Oblate spheroid

For the oblate spheroid shown in Figure 5.2b, we assume the short semi-axis "c" is aligned along z and that $a = b > c$. Then, from Equation (5.5b), $L_{\perp} = 1 - 2L_{\parallel}$. The elliptic integral used in the definition of $L_{\parallel}$ becomes

$$L_{\parallel} = \frac{a^2 c}{2}\int_0^{\infty}\frac{ds}{(s + a^2)^2\sqrt{s + c^2}} = \frac{a^2 c}{2(a^2 - c^2)}\left[\frac{\pi/2}{\sqrt{a^2 - c^2}} - \frac{c}{a^2}\right] \tag{5.20}$$

We may then combine Equation (5.20) with Equation (5.14) to evaluate the alignment torque for the oblate spheroid.

For a very thin oblate spheroid, $c \ll a = b$, so that $L_{\parallel} \ll 1$ and $L_{\perp} \approx 1$. Now the torque expression simplifies to

$$T^e_{\text{oblate}} = \frac{4\pi a^2 c}{3}\varepsilon_1 E_{\parallel}E_{\perp}\frac{(\varepsilon_2 - \varepsilon_1)^2}{[\varepsilon_1 + (\varepsilon_2 - \varepsilon_1)L_{\parallel}]\varepsilon_2} \tag{5.21}$$

The term involving $L_{\parallel} \ll 1$ is retained again, permitting us to use the $\varepsilon_2/\varepsilon_1 \to \infty$ limit to obtain a torque expression for a conducting oblate spheroid. A useful expression for $L_{\parallel}$ here is

$$L_{\parallel} \approx \frac{c}{2a}\left[\frac{\pi}{2} - \frac{c}{a}\right] \tag{5.22}$$

D. *DEP force on ellipsoidal particle*

In Chapter 3 concerning dielectrophoresis, only spherical particles were considered. For the sake of completeness, therefore, we consider the DEP force on a homogeneous and isotropic dielectric particle of ellipsoidal shape. Imagine the imposed electric field $\bar{E}_0$ on the ellipsoid to be slightly nonuniform. Consistent with the dielectrophoretic approximation (c.f., Section 2.1B), we calculate the effective dipole moment $\bar{p}_{\text{eff}}$ using the imposed electric field value at the center of the particle and then insert this expression into the general DEP force equation, Equation (2.4).

$$\bar{F}_{\text{DEP}} = \frac{4\pi abc\,(\varepsilon_2 - \varepsilon_1)}{3}\left[\frac{E_{0,x}}{1 + \left(\dfrac{\varepsilon_2 - \varepsilon_1}{\varepsilon_1}\right)L_x}\frac{\partial}{\partial x} + \ldots\right]\bar{E}_0 \qquad (5.23)$$

Shape-induced anisotropy makes the DEP force expression more complicated for an ellipsoidal particle than it is for a sphere. Even so, the force remains proportional to the scalar product of particle volume and excess permittivity, that is, $(\varepsilon_2 - \varepsilon_1)4\pi abc/3$. Ellipsoidal particles exhibit the same positive and negative DEP phenomenology, depending on whether ε_2 is greater than or less than ε_1, respectively.

For the special case of a prolate spheroidal particle $(b = c)$ aligned with its long axis parallel to and positioned on the axis of an axially symmetric field, the DEP force expression resembles that for a sphere.

$$\bar{F}_{\text{DEP}} = \frac{2\pi ab^2}{3}\left[\frac{(\varepsilon_2 - \varepsilon_1)}{1 + \left(\dfrac{\varepsilon_2 - \varepsilon_1}{\varepsilon_1}\right)L_{\|}}\right]\frac{\partial E_{0,z}^2}{\partial z}\hat{z} \qquad (5.24)$$

This expression is also valid for oblate spheroids $(a < b = c)$, though it should be noted that such a particle with its short axis parallel to the electric field will be in unstable alignment.

A useful expansion of the depolarization factor $L_{\|}$ in terms of $\gamma \equiv a/b$ is (Jones and Bliss, 1977)

$$L_{\|} = \left[1 + \frac{3}{5}(1 - \gamma^{-2}) + \frac{3}{7}(1 - \gamma^{-2})^2 + \ldots\right]/(3\gamma^{-2}) \qquad (5.25)$$

For a spherical particle, $\gamma = 1$ and $L_{\|} = 1/3$, in which case Equation (5.24) reduces to the familiar expression for the DEP force on a sphere.

5.3 Orientation for lossy dielectric ellipsoids

Orientational behavior becomes much more interesting for a dielectric ellipsoid with ohmic conductivity in a linearly polarized AC electric field because the preferred (or stable) orientation depends upon the frequency. Such *orientational spectra* are marked by a set of critical frequencies where the particle abruptly changes from one stable orientation to another. The explanation of this behavior is that Maxwell–Wagner polarization is characterized by a different relaxation time constant along each of the three principal axes. Many nonspherical particles and biological cells, including avian and mammalian erythrocytes (red blood cells), exhibit orientational spectra.

A. *Theory for homogeneous particles*

The analysis summarized below parallels earlier work for homogeneous spheroids (Gruzdev, 1965; Schwarz et al., 1965; Saito et al., 1966) that was extended to the case of ellipsoids (Miller and Jones, 1987). Assume that the ellipsoidal particle shown in Figure 5.1 and the medium in which it is suspended have conductivities σ_2 and σ_1, respectively. Furthermore, assume that the electric field vector is linearly polarized, with magnitude $\bar{E}_0$ and radian frequency ω. The electric field frequency is sufficiently high so that the particle responds only to the time-average torque. First, it is convenient to rewrite in more general form the effective moment components using Equation (5.3).

$$(\underline{p}_{\text{eff}})_\alpha = 4\pi abc\varepsilon_1 \underline{K}_\alpha \underline{E}_{0,\alpha} \tag{5.26}$$

where the index α represents x, y, or z.[3] Equation (5.26) introduces a generalization of the complex polarization factor $\underline{K}(\omega)$ employed in Chapters 3 and 4.

$$\underline{K}_\alpha \equiv \frac{\underline{\varepsilon}_2 - \underline{\varepsilon}_1}{3[\underline{\varepsilon}_1 + (\underline{\varepsilon}_2 - \underline{\varepsilon}_1)L_\alpha]}, \quad \alpha = x, y, \text{ or } z \tag{5.27}$$

where $\underline{\varepsilon}_1$ and $\underline{\varepsilon}_2$ are the same complex permittivities defined in Section 2.3A. If the particle is a sphere, that is, if $a = b = c$, then $L_x = L_y = L_z = 1/3$ and all $\underline{K}_\alpha$ revert to the form of Equation (2.30). For an ellipsoid, however, $\underline{K}_x$, $\underline{K}_y$, and $\underline{K}_z$ yield three distinct Maxwell–Wagner relaxation time constants, one for each of the principal axes.

$$(\tau_{\text{MW}})_\alpha = \frac{(1 - L_\alpha)\varepsilon_1 + L_\alpha\varepsilon_2}{(1 - L_\alpha)\sigma_1 + L_\alpha\sigma_2}, \quad \alpha = x, y, \text{ or } z \tag{5.28}$$

---

[3] Note that ε_1 and not $\underline{\varepsilon}_1$ appears in Equation (5.26). Refer to Section 2.3A for discussion of this point.

These time constants quantify the buildup or decay of free interfacial electric charge in response to the orthogonal field components E_x, E_y, and E_z.

We may now employ the effective moment method to derive expressions for the net torque on an ellipsoid in an AC electric field. Invoking the approximation that the electric field frequency is sufficiently high so that the particle responds only to time-averages, the α component of the alignment torque is

$$\langle T^e \rangle_\alpha = \frac{1}{2} \mathrm{Re} \left[(p_{\mathrm{eff}})_\beta E_{0,\gamma}^* - (p_{\mathrm{eff}})_\gamma E_{0,\beta}^* \right] \tag{5.29}$$

Here, the subscripts α, β, and γ are ordered according to the convention for a right-handed coordinate system, that is, $x \to y \to z \to x$. Equations (5.26) and (5.29) combine to yield expressions for the torque components along any of the three axes.

$$\langle T^e \rangle_\alpha = \frac{2}{3} \pi a b c \varepsilon_1 (L_\gamma - L_\beta) E_{0,\beta} E_{0,\gamma} \mathrm{Re}[\underline{K}_\beta \underline{K}_\gamma] \tag{5.30}$$

Equation (5.30) is the frequency-dependent generalization of Equation (5.8).

B. Alignment behavior of homogeneous particles

Just as in the relations (5.9a,b,c), the signs of the x, y, and z components of the time-average electrical torque depend on the relative magnitudes of the depolarization factors L_x, L_y, and L_z, but they now also depend on the frequency ω through the terms $\mathrm{Re}[K_x K_y]$, etc. To initiate a systematic investigation of the frequency-dependent behavior, we assume, once again, that $a > b > c$, so that $0 < L_x < L_y < L_z < 1$. Table 5.1 summarizes the preferred orientation of the ellipsoid according to the sign of $\mathrm{Re}[\underline{K}_\beta \underline{K}_\gamma]$, indicating the predicted axis of orientation – a, b, or c – for all possible combinations of signs for the torque components. The two cases where no preferred orientation is predicted can be disallowed on mathematical grounds; thus, there always exists a single stable orientation for an ellipsoidal particle in an AC electric field.

The signs of the torque components depend on the frequency ω. As each critical frequency is encountered, the particle's existing orientation becomes unstable and it turns over abruptly, seeking a new stable orientation along a different axis. The *orientational spectrum* consists of all the turnover frequencies plus the preferred orientation within each frequency band defined by these turnovers. Table 5.1 can be used to predict the stable orientation of a model particle at any frequency.

One can investigate the frequency dependence of the $\mathrm{Re}[\underline{K}_\beta \underline{K}_\gamma]$ terms in the three torque components using a set of three biquadratic equations. This

Table 5.1 *Relationship of torque signs to preferred (stable) axis of orientation with respect to the electric field vector for dielectric ohmic ellipsoid suspended in dielectric ohmic liquid*

$<T^e>_x \propto -\mathrm{Re}[\underline{K}_y\underline{K}_z]$	$<T^e>_y \propto -\mathrm{Re}[\underline{K}_z\underline{K}_x]$	$<T^e>_z \propto -\mathrm{Re}[\underline{K}_x\underline{K}_y]$	Orientation
+	+	+	None
+	+	−	b
+	−	+	a
+	−	−	b
−	+	+	c
−	+	−	c
−	−	+	a
−	−	−	None

Source: Miller, 1989.

approach yields a systematic set of rules for predicting the frequency-dependent turnovers of a homogeneous dielectric ellipsoid with ohmic conductivity (Miller, 1989). Given that $a > b > c$, the turnovers must occur according to one of the sequences summarized in Table 5.2. Keep in mind that there is only one stable orientation at each frequency. Furthermore, at the lowest and highest frequencies, the "a" orientation – that with the longest axis parallel to the electric field – is always preferred. The existence of other stable orientations at intermediate frequencies depends upon the relative values of $\varepsilon_2/\varepsilon_1$ and σ_2/σ_1 as well as the ratios of the three semi-axes: a, b, and c. Frequency-dependent turnover behavior is guaranteed if either (i) $\varepsilon_2/\varepsilon_1 < 1 < \sigma_2/\sigma_1$ or (ii) $\sigma_2/\sigma_1 < 1 < \varepsilon_2/\varepsilon_1$. Outside of these limits, turnovers, while still possible, may not exist.

By plotting the boundaries between different orientation regimes, we can observe how the relative dielectric permittivities $\varepsilon_2/\varepsilon_1$ and conductivities σ_2/σ_1 influence frequency-dependent alignment. Refer to Figures 5.3a and b. In these plots, each curve represents the marginal condition where one torque component is zero.

C. Orientation of layered particles

Certain biological cells, such as avian and mammalian erythrocytes, feature pronounced ellipsoidal shape and consequently exhibit frequency-dependent orientation in a linearly polarized electric field. Erythrocytes consist of a relatively

Table 5.2 *Orientation and allowable turnover sequences going from low to high frequencies for a homogeneous dielectric ellipsoid with ohmic losses and semi-axes a > b > c*

	$\sigma_2 > \sigma_1$	$\sigma_2 < \sigma_1$
$\varepsilon_2 > \varepsilon_1$	$a, a{\rightarrow}b{\rightarrow}a$, or $a{\rightarrow}b{\rightarrow}c{\rightarrow}b{\rightarrow}a$ (+DEP @ all freq.)	$a{\rightarrow}c{\rightarrow}b{\rightarrow}a$ (−DEP @ low freq. & +DEP @ high freq.)
$\varepsilon_2 < \varepsilon_1$	$a{\rightarrow}b{\rightarrow}c{\rightarrow}a$ (+DEP @ low freq. & −DEP @ high freq.)	a or $a{\rightarrow}c{\rightarrow}a$ (−DEP @ all freq.)

Note: For reference, the positive and negative DEP regimes for spheres are indicated in parentheses. See Table 3.1.
Source: Adapted from Miller, 1989.

conductive cytoplasmic interior (with or without a nucleus) enclosed by a thin, selectively permeable membrane characterized by uniform capacitance per unit area c_m. There are also examples of artificial particles having ellipsoidal shape and surface treatments best represented by a uniform surface layer. Attempts to model ellipsoids with uniform layers must deal with certain theoretical difficulties. In contrast to the spherical shells of Appendix C, no simple replacement of a layered ellipsoid by an equivalent homogeneous and isotropic ellipsoid is possible. This inconvenience is the inescapable consequence of the shape-dependent anisotropy. The uniform layer introduces a term into the boundary condition at the surface that cannot be expressed in closed form using ellipsoidal harmonics.

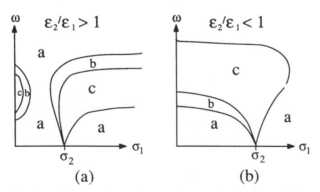

(a)   (b)

Fig. 5.3 Map of stable orientations for the homogeneous ellipsoid shown in Figure 5.1. The curves delineating the different regions, obtained by setting individual torque components to zero, are only representative. The existence of multiple turnover frequencies is assured when either (a) $\varepsilon_2/\varepsilon_1 > 1 > \sigma_2/\sigma_1$ or (b) $\varepsilon_2/\varepsilon_1 < 1 < \sigma_2/\sigma_1$. Outside these limits, the existence of turnovers is not guaranteed.

Success in calculating the orientational torque for layered ellipsoidal particles is possible if one assumes that all layers are defined by confocal surfaces (Stepin, 1965; Asami et al., 1980). As shown in Figure 5.4a, confocal layers are not uniform over the surface of the particle. When the confocal model is employed, the equivalent ellipsoid used to predict orientational spectra is obtained by the substitution of a homogeneous but *anisotropic* permittivity into the expressions for the various torque components. For an ellipsoid of complex permittivity ε_2 covered by a very thin (confocal) layer having real permittivity ε_m, the x component of ε_2' becomes, for example,

$$(\varepsilon_2)_x' = \varepsilon_m \left[\frac{\varepsilon_2 + \Delta_x (\varepsilon_2 + \varepsilon_m)/a}{\varepsilon_m + \Delta_x (\varepsilon_2 + \varepsilon_m)/a} \right] \tag{5.31}$$

where Δ_x is the thickness of the layer measured along the x axis. Given Δ_y and Δ_z, similar expressions may be written for $(\varepsilon_2)_y'$ and $(\varepsilon_2)_z'$, respectively.

To evaluate the torque components and thus predict orientational spectra, it is not correct to substitute these expressions for the permittivity directly into Equation (5.27) because the expressions for K_α are based in the assumption of isotropy. Instead, we must use $(\varepsilon_2)_x'$, $(\varepsilon_2)_y'$, and $(\varepsilon_2)_z'$ directly in the expressions for the components of the effective moment p_{eff} and then evaluate the time-average torque components using $(1/2)\mathrm{Re}[\bar{p}_{eff} \times \bar{E}_0^*]$. This approach to the prediction of orientational spectra is tedious because, in general, the turnover

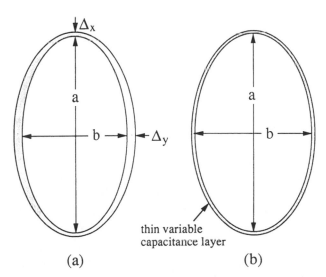

thin variable
capacitance layer

(a)   (b)

Fig. 5.4 Two simple approximate models for an ellipsoidal particle with a uniform surface layer. (a) Layer defined by confocal ellipsoidal surfaces. (b) Very thin layer with variable surface capacitance.

criteria are defined in terms of high-order polynomials that cannot be factored readily, if at all.

A somewhat simpler alternative to the confocal model is to place a spatially varying capacitance at the surface of the ellipsoid (Miller, 1989). Refer to Figure 5.4b. Results obtained with the spatially varying capacitance model differ only marginally from those obtained using a confocal representation.

Aside from the increased mathematical complexity attributed to the anisotropy of the equivalent particle for these particles, no real conceptual difficulties arise in prediction of the orientational spectra for layered ellipsoids. Layers only have the effect of introducing additional Maxwell–Wagner relaxation frequencies and introducing the possibility of additional turnovers. The confocal and variable surface capacitance models (Figure 5.4a and b) are not really equivalent to an ellipsoid with a *uniform* surface layer; however, comparison of the results obtained with them does reveal that the predicted orientational spectra are relatively insensitive to the choice of models.

5.4 Experimental orientational spectra

Measuring the dielectric properties of ellipsoidal particles using conventional three-terminal bridge techniques and mixture formulas is unsatisfactory because it is not possible to know with any certainty the orientation of the cells as the frequency is adjusted. Precision measurements using DEP levitation are complicated because, within the highly nonuniform electric field of the levitator, elongated cells sometimes exhibit skewed orientations with respect to the axial electric field. Therefore, orientational spectra fill an important gap by providing the means to study the properties of elongated particles and cells, such as erythrocytes.

Figure 5.5 presents orientational data obtained over the frequency range from ~1 kHz to ~400 kHz for small particles of titanium dioxide (TiO_2) suspended in isopropanol (Miller, 1989). The particles, approximately 1 mm in their longest dimension with semi-axis ratio of $a{:}b{:}c \approx 4{:}2{:}1$ were only crudely ellipsoidal. The electric field strength was maintained at ~2 kV/cm, except at the highest frequencies, where the amplifier output dropped off due to an impedance mismatch. By dissolving varied amounts of calcium chloride ($CaCl_2$) in the isopropanol, the solution conductivity σ_1 could be adjusted between ~20 and ~180 μS/m. The data points plotted in Figure 5.5 signify the dominant orientation observed with an optical microscope for each set of frequency and conductivity values. The theoretical curves denote the predicted turnover conditions where individual torque components go through zero. Despite certain limitations, the data largely verify the turnover sequence $a{\to}c{\to}b{\to}a$ predicted for these inorganic particles.

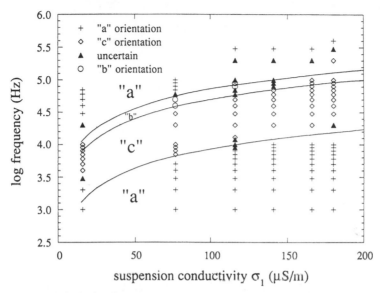

Fig. 5.5 Orientational spectra of titanium dioxide particles suspended in isopropanol doped with calcium chloride (from Miller, 1989). The theoretical curves represent turnover conditions obtained by setting various torque components to zero for $a:b:c = 4:2:1$; $\varepsilon_2 = 100\,\varepsilon_0$; $\sigma_2 = 0.0$ S/m; $\varepsilon_1 = 30\,\varepsilon_0$.

Orientational spectra for red blood cells provide a test of the theory summarized in Section 5.3C. Figure 5.6 displays data obtained with llama erythrocytes, which are noted for their pronounced ellipsoidal shape, namely, $a:b:c \approx$ 4.0:2.0:1.1 (± 0.2) μm and which provide the opportunity to observe all three orientations (Miller and Jones, 1993). Some transitions are indistinct and the predicted "*b*" orientation at higher frequencies is not observed at all. Still, the main features of the theoretical predictions are borne out by the data. Miller used these experimental orientational spectra along with a variant of the confocal ellipsoid model to estimate the cytoplasmic conductivity at $\sigma_2 = 0.26$ S/m, a value consistent with published data obtained by more conventional means with erythrocytes of more nearly spherical shape.

5.5 Static torque on suspended particle

Recall from Chapter 4 that a rotational torque, due to dielectric or ohmic loss, induces continuous rotation of spherical particles. On the other hand, from Section 5.3, we have learned that particles with shape anisotropy and loss experience a frequency-dependent orientational torque. From the standpoint of experimental observations made on a suspended particle, both the rotational and orientational effects are likely to be present at the same time. Therefore, it is important to

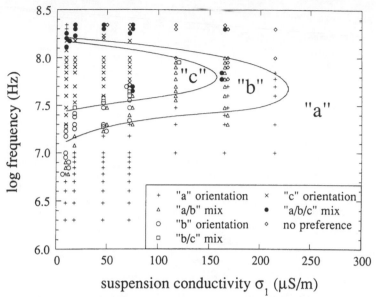

Fig. 5.6 Orientational spectra of llama erythrocytes suspended in isotonic sugar/calcium chloride solutions (from Miller, 1989). The theoretical curves represent turnover conditions defined by setting various torque components to zero using $a:b:c = 4:2:1.1$; $\varepsilon_2 = 52\,\varepsilon_0$; $\sigma_2 = 0.23$ S/m; $\varepsilon_m = 10\,\varepsilon_0$; $\sigma_m = 0.0$ S/m; $\varepsilon_1 = 80\,\varepsilon_0$.

investigate their combined influence upon the behavior of a nonspherical particle. The analysis that follows, based upon that of Ogawa (1961), is particularly instructive because it reveals that rotational and orientational torques are sensitive to different dielectric properties of the particle and suspension medium.

A. Basic model for torque calculation

Consider an ellipsoidal particle with semi-axes a, b, and c suspended vertically by a fine thread or fiber along its "c" axis in some medium of permittivity ε_1 and conductivity σ_1 within an electrode chamber. Assume that the electrodes can be excited by multiphase AC voltage to impose a fairly uniform electric field of arbitrary polarization. Figure 5.7 shows a top view of the arrangement with the cross-section of a suspended particle. With means provided to measure the electromechanical torque exerted upon the particle, this simple arrangement becomes an instrument to measure the complex polarization of small fibers, crystals, and other anisotropic particles. The particle is suspended by a thread or fiber so that its rotation is restrained by torsion.

For convenience in this discussion, we employ two rectangular coordinate systems: x, y, z (corresponding to the three principal axes of the particle) and x', y', z' (corresponding to the electrode axes). Note that z and z' are identical. The angle

Top View

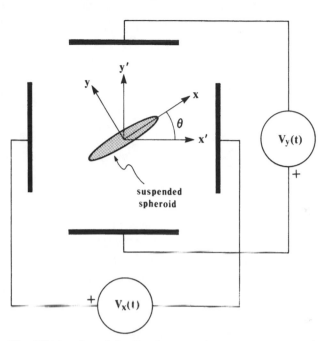

Fig. 5.7 Top view of chamber for measuring the torque on nonspherical particles, fibers, etc., in an AC electric field. The "*a*" and "*b*" semi-axes of the particle are aligned with the *x* and *y* coordinates, respectively, and the particle is free to rotate only in the *x–y* plane. Proper AC phasing of $V_x(t)$ and $V_y(t)$ makes it possible to create either a linearly polarized or a rotating electric field.

between the x and x' axes is defined as θ. The electric field components in the coordinate system of the particle expressed in terms of the axes of the electrodes are

$$\underline{E}_x = \underline{E}_{x'}\cos\theta + \underline{E}_{y'}\sin\theta \tag{5.32a}$$

$$\underline{E}_y = -\underline{E}_{x'}\sin\theta + \underline{E}_{y'}\cos\theta \tag{5.32b}$$

Recognition of material anisotropy is vital here because the types of particles one might envision investigating with the apparatus of Figure 5.7 include crystals and fibers, which are likely to exhibit strong dielectric anisotropy (Ogawa, 1961). The complex permittivity tensor for the suspended particle is

$$\bar{\bar{\underline{\varepsilon}}}_2 \equiv \begin{bmatrix} \varepsilon_x & 0 & 0 \\ 0 & \varepsilon_y & 0 \\ 0 & 0 & \varepsilon_z \end{bmatrix} \tag{5.33}$$

where $\underline{\varepsilon}_x = \varepsilon_x + \sigma_x/j\omega$, etc. In Equation (5.33) we have assumed that all off-diagonal terms are zero, which is true only if the semi-axes of the particle correspond to the principal (crystalline) axes of the material.

Using Equations (5.26) and (5.29), an expression for the time-average torque about the z axis may be written in terms of $\underline{E}_x$ and $\underline{E}_y$.

$$\langle T_z^e \rangle = \frac{2\pi abc}{3}\varepsilon_1 \mathrm{Re}\left[\frac{\underline{\varepsilon}_x - \underline{\varepsilon}_1}{[\underline{\varepsilon}_1 + (\underline{\varepsilon}_x - \underline{\varepsilon}_1)L_x]}\underline{E}_x\underline{E}_y^* - \frac{\underline{\varepsilon}_y - \underline{\varepsilon}_1}{[\underline{\varepsilon}_1 + (\underline{\varepsilon}_y - \underline{\varepsilon}_1)L_y]}\underline{E}_y\underline{E}_x^* \right] \quad (5.34)$$

where L_x and L_y are defined by Equation (5.4). Equation (5.34) is a general result, valid for any anisotropic, ellipsoidal particle as long as particle permittivity takes the form of Equation (5.33). One finds it easier to distinguish between rotational and orientational torque effects on a suspended particle by considering separately the cases of circularly and linearly polarized electric fields.

For the present, we limit attention to spherical, oblate, and prolate spheroids having material isotropy, that is, $\underline{\varepsilon}_x = \underline{\varepsilon}_y = \underline{\varepsilon}_z \rightarrow \underline{\varepsilon}_2$ (Ogawa, 1961). Effects of crystalline anisotropy upon particle alignment phenomena, of equal or greater importance with respect to magnetic materials, are examined more thoroughly in Section 5.6.

B. Rotational torque on suspended particle

To investigate the rotational torque, we start with a CCW rotating electric field.

$$(\underline{E}_{x'} = E_0) \quad \text{and} \quad \underline{E}_{y'} = -jE_0 \tag{5.35}$$

Then, for an isotropic ellipsoidal particle, Equation (5.34) takes the form

$$\langle T_z^e \rangle = -\frac{2\pi abc}{3}\varepsilon_1 E_0^2 \mathrm{Im}\left[\frac{(\underline{\varepsilon}_2 - \underline{\varepsilon}_1)[2\underline{\varepsilon}_1 + (\underline{\varepsilon}_2 - \underline{\varepsilon}_1)(L_x + L_y)]}{[\underline{\varepsilon}_1 + (\underline{\varepsilon}_2 - \underline{\varepsilon}_1)L_x][\underline{\varepsilon}_1 + (\underline{\varepsilon}_2 - \underline{\varepsilon}_1)L_y]} \right] \tag{5.36}$$

As a validation of Equation (5.36), consider the sphere, that is: $a = b = c = R$ and $L_x = L_y = L_z = 1/3$. Here, the time-average torque reverts to familiar form.

$$\langle T_z^e \rangle = -4\pi R^3 \varepsilon_1 E_0^2 \mathrm{Im}\left[\frac{\underline{\varepsilon}_2 - \underline{\varepsilon}_1}{\underline{\varepsilon}_2 + 2\underline{\varepsilon}_1} \right] \tag{5.37}$$

which is identical to Equation (2.50).

B.1 Oblate spheroid

Consider a very thin oblate spheroid suspended with its short axis c aligned along the torsional axis z, as shown in Figure 5.8a. For this case, $a = b \gg c$ and $L_x = L_y \ll 1$, $L_z \approx 1$. Using these approximations in Equation (5.36), we obtain

$$\langle T_z^e \rangle = -\frac{4\pi a^2 c}{3} \varepsilon_1 E_0^2 \, \mathrm{Im}\left[\frac{\varepsilon_2 - \varepsilon_1}{\varepsilon_1}\right]$$

$$= -\frac{4\pi a^2 c}{3} \varepsilon_1 E_0^2 \left(\frac{\varepsilon_2}{\varepsilon_1}\right)\left[\frac{(1 - \tau_1/\tau_2)\,\omega\tau_1}{1 + (\omega\tau_1)^2}\right]$$

(5.38)

The frequency dependence of the rotational torque on the oblate spheroid is the same as that for the sphere, Equation (5.37), except that the maximum rotational torque is achieved at an electric field frequency of $\omega = 1/\tau_1$, rather than $1/\tau_{MW}$. This is because the thin disk, when aligned parallel to the x–y plane, has a very small depolarization effect on the imposed electric field.

B.2 Prolate spheroid

Another case for which the rotational torque may be determined easily is the highly elongated prolate spheroid with its longest axis parallel to the z axis, that is, $c \gg a = b$, so that $L_x = L_y \approx 1/2$ and $L_z \ll 1$. Refer to Figure 5.8b. The time-average torque then becomes

$$\langle T_z^e \rangle = -\frac{8\pi a^2 c}{3} \varepsilon_1 E_0^2 \, \mathrm{Im}\left[\frac{\varepsilon_2 - \varepsilon_1}{\varepsilon_2 + \varepsilon_1}\right]$$

$$= -\frac{16\pi a^2 c}{3} \varepsilon_1 E_0^2 \left[\frac{(1 - \tau_1/\tau_2)\,\omega\tau_{MW}}{(1 + \sigma_2/\sigma_1)\,(1 + \varepsilon_1/\varepsilon_2)\,[1 + (\omega\tau_{MW})^2]}\right]$$

(5.39)

where τ_{MW} now departs slightly from the form of Equation (2.32).

$$\tau_{MW} = \frac{\varepsilon_2 + \varepsilon_1}{\sigma_2 + \sigma_1}$$

(5.40)

which is identical to the Maxwell–Wagner charge relaxation time of an infinite cylinder having its axis of rotation perpendicular to the imposed electric field (Melcher and Taylor, 1969).

B.3 Orientations exhibiting saliency

Figures 5.8c and d display two other orientations, each distinguished by the fact that the particle projection in the x'–y' plane is elliptical. This *saliency* introduces a pulsating component to the torque when the rotating field is applied, so that $T^e(t)$ will have both a smooth time-average component and a sinusoidal contribution. The sinusoidal term may cause the suspended particle to vibrate. If the particle exhibiting saliency in the x–y plane is free to rotate, then the electrical torque will induce time-average motion with a superimposed pulsating component. It is also possible for such a particle to rotate in *synchronism* with the electric field. To understand this behavior, imagine that the electric field is rotating very slowly so that the particle, trying to remain aligned with the field, rotates at the same speed but with some lag angle. If the electric field frequency is too high or if the viscous drag is too great, then such synchronism is unlikely but asynchronous rotation will still be possible.

C. Alignment torque on suspended particle

A linearly polarized electric field results if $\underline{E}_{x'}$ and $\underline{E}_{y'}$ are in phase. In contrast to Equation (5.35), let us now assume

$$\underline{E}_{x'} = \underline{E}_{y'} = E_0 \qquad\qquad (5.41)$$

which represents a linearly polarized electric field making a 45° angle with respect to the electrode axis x'. Now, Equation (5.34) becomes

$$\langle T_z^e \rangle = \frac{2\pi abc}{3}\, \varepsilon_1 E_0^2 (L_y - L_x)\mathrm{Re}\left[\frac{(\varepsilon_2 - \varepsilon_1)^2}{[\varepsilon_1 + (\varepsilon_2 - \varepsilon_1)L_x][\varepsilon_1 + (\varepsilon_2 - \varepsilon_1)L_y]}\right]\cos 2\theta$$

$$(5.42)$$

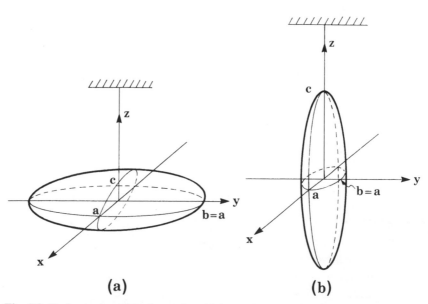

(a)   **(b)**

Fig. 5.8 Various orientations for a spheroidal particle suspended in electrode structure for torque measurements of dielectric properties. (a) Oblate spheroid with no x–y saliency: $a = b \gg c$. (b) Prolate spheroid with no x–y saliency: $c \gg a = b$. (c) Oblate spheroid with x–y saliency: $a = c \gg b$. (d) Prolate spheroid with x–y saliency: $a \gg b = c$. [Parts (c) and (d) are on facing page.]

As expected, this time-average torque depends on the angle θ and is thus classified as an alignment torque. Furthermore, its sign depends upon the relative values of the depolarization quantities L_x and L_y. A linearly polarized electric field exerts no torque on either a stationary sphere or on any spheroidal particle having a circular projection in the x'–y' plane.

It is useful to examine the frequency dependence of the time-average torque $\langle T_z^e \rangle$ for the fairly typical situation of a particle suspended in air, in which case

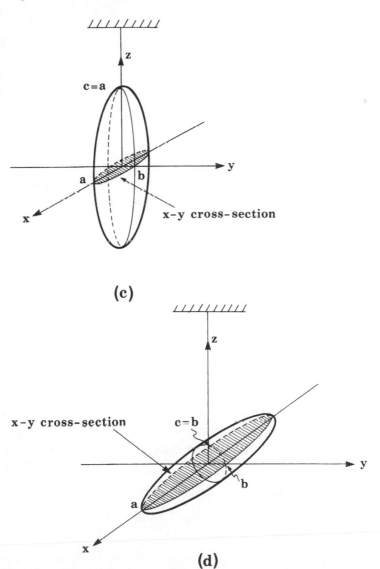

Fig. 5.8 Caption on facing page.

$\varepsilon_1 = \varepsilon_0$ and $\sigma_1 \approx 0$. The real part of the bracketed term in Equation (5.42) becomes

$$\mathrm{Re}\left[\frac{(\varepsilon_2 - \varepsilon_1)^2}{[\varepsilon_1 + (\varepsilon_2 - \varepsilon_1)L_x][\varepsilon_1 + (\varepsilon_2 - \varepsilon_1)L_y]}\right] \rightarrow \frac{1}{L_x L_y}\mathrm{Re}\left[\frac{[j\omega(\varepsilon_2 - \varepsilon_0)/\sigma_2 + 1]^2}{[j\omega\tau_x + 1][j\omega\tau_y + 1]}\right]$$

$$(5.43)$$

The time constants are given by Equation (5.28), so that

$$\tau_x = \frac{\varepsilon_0(1 - L_x) + \varepsilon_2 L_x}{\sigma_2 L_x} \quad \text{and} \quad \tau_y = \frac{\varepsilon_0(1 - L_y) + \varepsilon_2 L_y}{\sigma_2 L_y} \tag{5.44}$$

Of practical interest is the case of a prolate spheroid suspended with its long axis in the $x'–y'$ plane. Assume that $a \gg b = c$ so that $L_y = L_z \approx 1/2$ and $L_x \ll 1$, as shown in Figure 5.8d. Under these assumptions, the equations for the time constants may be simplified.

$$\tau_x \approx \frac{\varepsilon_0}{\sigma_2 L_x} \quad \text{and} \quad \tau_y \approx \frac{\varepsilon_0 + \varepsilon_2}{\sigma_2} \tag{5.45}$$

Here, note that $\tau_x \gg \tau_y$ so that the time-average torque starts to roll off from the DC limit at $\omega = \sigma_2 L_x/\varepsilon_0$. With Equation (5.17) used to approximate L_x, this critical frequency correlates rather well with the frequency-dependent torque data reported for long wood fibers containing moisture (Talbott and Stefanakos, 1972). Note that this and all other critical frequencies scale directly with particle conductivity σ_2.

5.6 Orientation of magnetizable particles

We may exploit the analogy between linear dielectric and magnetizable media to derive alignment torque expressions for magnetic ellipsoids. As expected, the static alignment behavior of isotropic magnetic ellipsoids is identical to dielectric particles; however, anisotropy plays an important role in the orientational behavior of many magnetic particles. Therefore, in this section we focus on orientational phenomena in anisotropic materials. In addition, the cases of isotropic particles with remanent magnetization and magnetic saturation receive some attention.[4]

A. *Magnetically linear particles with anisotropy*

Consider a homogeneous, magnetizable ellipsoid with semi-axes a, b, and c suspended in air (or vacuum) so that $\mu_1 = \mu_0$. Let the particle be anisotropic

---

[4] To the author's knowledge, the problem of frequency-dependent orientation for linear magnetic ellipsoids with induced eddy currents has not been worked.

with tensor magnetic susceptibility defined by the linear constitutive law $\overline{M}(\overline{H}) = \overline{\overline{\chi}} \bullet \overline{H}$, where $\overline{M}$ is the volume magnetization of the material and

$$
\overline{\overline{\chi}} = \begin{bmatrix} \chi_{xx} & \chi_{xy} & \chi_{xz} \\ \chi_{yx} & \chi_{yy} & \chi_{yz} \\ \chi_{zx} & \chi_{zy} & \chi_{zz} \end{bmatrix} \tag{5.46}
$$

The reciprocity condition dictates that $\chi_{xy} = \chi_{yx}$, $\chi_{yz} = \chi_{zy}$, and $\chi_{zx} = \chi_{xz}$. The effective moment vector $\overline{m}_{\text{eff}}$ for the uniformly magnetized particle is obtained by multiplying $\overline{M}$ by the particle volume, that is, $V = 4\pi abc/3$.

$$
\overline{m}_{\text{eff}} = V\overline{M}(\overline{H}^-) \tag{5.47}
$$

$\overline{H}^-$ is the magnetic intensity within the particle. The magnetic alignment torque is computed using the magnetic torque expression $\overline{T}^m = \mu_0(\overline{m}_{\text{eff}} \times \overline{H})$ taken from Table A.2 in Appendix A.

Consider the case when the principal crystalline axes correspond to the semi-axes of the ellipsoidal particle, that is, $\chi_{xy} = \chi_{yz} = \chi_{zx} = 0$. In this limit, we realize considerable simplification in calculating the alignment torque. Assume that the ellipsoidal particle, aligned with its semi-axes (and principal axes) parallel to x, y, and z, is subjected to a uniform magnetic intensity vector $\overline{H}_0$, where

$$
\overline{H}_0 = \hat{x}H_{0,x} + \hat{y}H_{0,y} + \hat{z}H_{0,z} \tag{5.48}
$$

The three components of $\overline{H}_0$ may be expressed in terms of its magnitude $H_0 = |\overline{H}_0|$ and the direction cosines c_x, c_y, and c_z.

$$
H_{0,x} = c_x H_0, \; H_{0,y} = c_y H_0, \; \& \; H_{0,z} = c_z H_0 \tag{5.49}
$$

Without loss of generality, we may specify that c_x, c_y, and c_z are all positive. Then, the magnetic intensity vector within the particle is

$$
\overline{H}_0^- = \hat{x}\frac{c_x H_0}{1 + \chi_{xx}L_x} + \hat{y}\frac{c_y H_0}{1 + \chi_{yy}L_y} + \hat{z}\frac{c_z H_0}{1 + \chi_{zz}L_z} \tag{5.50}
$$

The magnetic torque components acting upon the ellipsoid are all of similar form, that is

$$
T_\gamma^m = \frac{4\pi abc}{3}\mu_0 H_0^2 \left[\frac{\chi_{\alpha\alpha}}{1 + \chi_{\alpha\alpha}L_\alpha} - \frac{\chi_{\beta\beta}}{1 + \chi_{\beta\beta}L_\beta} \right] c_\alpha c_\beta \tag{5.51}
$$

As before, the indices α, β, and γ represent an ordered sequence of right-hand coordinates, that is, x, y, z. We may rewrite Equation (5.51) in a different form to highlight the distinctions between the effects of shape- and material-dependent anisotropy upon particle orientation.

$$T^m_\gamma = \frac{4\pi abc}{3}\mu_0 H_0^2 \left[\frac{\chi_{\alpha\alpha} - \chi_{\beta\beta} + \chi_{\alpha\alpha}\chi_{\beta\beta}(L_\beta - L_\alpha)}{(1 + \chi_{\alpha\alpha}L_\alpha)(1 + \chi_{\beta\beta}L_\beta)} \right] c_\alpha c_\beta \qquad (5.52)$$

Several limiting cases of Equation (5.52) can be checked easily. For example, in the limit of material isotropy, that is, $\chi_{xx} = \chi_{yy} = \chi_{zz}$, the particle will orient with its longest axis parallel to the imposed magnetic intensity vector. Furthermore, a spherical particle ($L_x = L_y = L_z$) with material anisotropy will align itself with the magnetic field parallel to the axis having the highest susceptibility.

In general, however, crystalline- and shape-dependent anisotropy will compete to control the particle's orientation. We can develop an appreciation for this competition by considering the special cases of prolate and oblate spheroidal crystals.

B. Prolate spheroidal crystals

The prolate spheroid serves as an easy-to-use model for a rod- or needle-shaped particle, so let us consider such a particle with $a > b = c$ and $\chi_{yy} = \chi_{zz}$. It is convenient to distinguish quantities either parallel ($\parallel$) or perpendicular ($\perp$) to the long particle axis: $L_\parallel \equiv L_x$ and $L_\perp \equiv L_y = L_z$, where $0 < L_\parallel < L_\perp$; $\chi_\parallel \equiv \chi_{xx}$ and $\chi_\perp \equiv \chi_{yy} = \chi_{zz}$. From Equation (5.52), the x component of torque is zero, while

$$T^m_y = -T^m_z = \frac{4\pi ab^2}{3}\mu_0 H_0^2 \left[\frac{\chi_\perp - \chi_\parallel + \chi_\perp\chi_\parallel(L_\parallel - L_\perp)}{(1 + \chi_\perp L_\perp)(1 + \chi_\parallel L_\parallel)} \right] c_y c_z \qquad (5.53)$$

The direction cosines c_y and c_z are both positive. Therefore, the torque signs are determined by the numerator of the bracketed term in Equation (5.53). By making reference to Table 5.1 once again and taking the condition $T^m_x = 0$ into account, we establish stable alignments for the spheroid with the following set of inequalities.

$\chi_\parallel > \chi^*$, orientation parallel to "a" axis $\qquad\qquad\qquad\qquad (5.54)$

$\chi_\parallel < \chi^*$, orientation perpendicular to "a" axis[5] $\qquad\qquad\qquad (5.55)$

---

[5] Orientation in the equatorial plane will be indifferent unless $b \neq c$, in which case particle alignment favors the longer of the two axes.

where $\chi^* \equiv \chi_\perp / [1 + \chi_\perp (L_\perp - L_{||})]$. The competition between material and shape anisotropy to establish stable orientation is apparent from these relations. Either parallel or perpendicular alignment is possible, depending upon the relative magnitudes of $\chi_{||}$ and χ^*.

C. Thin disks and laminae

The oblate spheroidal shape provides a convenient representation for thin disks and laminae. In this case, $a = b >> c$ so that $L_x = L_y << 1$ and $L_z \approx 1$. We again assume that x, y, and z correspond to the principal crystalline axes, but now let $\chi_{xx} \neq \chi_{yy}$. Furthermore, we restrict the imposed magnetic intensity vector to the x–y plane, that is

$$\bar{H}_0 = \hat{x}H_0 \cos\theta + \hat{y}H_0 \sin\theta \tag{5.56}$$

The direction cosines of Equation (5.49) now become $c_x = \cos\theta$ and $c_y = \sin\theta$, with $c_z = 0$. The only nonzero torque term is the z component.

$$T_z^m = \frac{2\pi a^2 c}{3} \mu_0 H_0^2 (\chi_{xx} - \chi_{yy}) \sin 2\theta \tag{5.57}$$

This expression is consistent with Eq. (96) in Chap. 4 of Bates (1961) for the torque exerted upon a thin disk having radius a and thickness $2c$. Here, orientation is determined solely by material anisotropy.

D. Isotropic particle with remanent magnetization

Particles composed of hard ferromagnetic material exhibit remanent magnetism, and often exhibit a permanent magnetic moment: $\bar{m}_{\text{perm}} = V\bar{M}_{\text{rem}}$ where $\bar{M}_{\text{rem}}$ is the remanent magnetization per unit volume defined in Figure 3.9 and V is the particle volume. Just like a compass needle, such a particle suspended in air will tend to align itself parallel to the applied magnetic field, even if the particle is spherical. The expression for the alignment torque is

$$T^m = \mu_0 V M_{\text{rem}} H_0 \sin\theta \tag{5.58}$$

where θ is the angle between the permanent moment and the applied magnetic field vectors. Equation (5.57) is a valid approximation as long as the applied field H_0 induces insignificant additional magnetization within the particle.

In the presence of a rotating magnetic field, a permanently magnetized particle can rotate steadily only at the synchronous condition $\Omega = \omega$. This case is clearly differentiated from the hysteresis effect described in Section 4.5B, where the torque is constant at all rotational speeds up to synchronism.

5.7 Applications of orientational phenomena

A. Dielectric particles

Several investigators have exploited static electromechanical torque measurement methods to study the AC dielectric properties of elongated particles. For example, Kawai and Marutake (1948) measured dielectric dispersion in long, thin crystals of Rochelle salt for varied temperature and humidity from 0.25 Hz to 1 kHz. The crystals, suspended on glass fibers, were ~30 mm in length and ~1 mm in width. More recently, Ogawa used the same technique to investigate dispersion in rod- and disk-shaped samples of photosensitive materials (Ogawa, 1961). As described in Section 5.5, he employed a rotating field to obtain $\varepsilon''(\omega)$ and a linearly polarized field to obtain $\varepsilon'(\omega)$.

Another application of virtually the same technique involved static torque measurements on conductive wood fibers using DC and 20-Hz to 1-kHz AC electric fields (Talbott and Stefanakos, 1972). Talbott and Stefanakos discovered that humidity has a strong influence upon the frequency dependence of the torque. The practical goal of this research was the application of electric fields to align long wooden fibers for the production of high-strength, anisotropic particle board. A more recent but very similar application of electric field–induced particle orientation exploits the effect to produce ceramic fiber–reinforced metal composites (Masuda and Itoh, 1989). Similar to wood fibers, the orientation of ceramic fibers is strongly influenced by surface conductivity. The time-dependent orientation of ceramic fibers suspended in dielectric liquids has been investigated in relation to this application (Itoh et al., 1994).

The red blood cells (erythrocytes) of warm-blooded animals vary considerably in their shape from specie to specie. Human erythrocytes, which are somewhat flattened and resemble oblate spheroids, manifest frequency-dependent orientational behavior (Griffin, 1970; Miller, 1989). In fact, it is found that the orientational spectra of cells drawn from sickle cell anemia patients differ from healthy cells (Vienken et al., 1984). The red blood cells of birds and ruminant mammals (camels, llamas, etc.) are often quite elongated and, as already discussed in Section 5.4, can exhibit three different orientations. Such spectra have been used to estimate cell parameters such as membrane capacitance and cytoplasmic conductivity (Miller, 1989).

Biomedical applications for electric field orientation extends beyond cells to other very important bioparticles. For example, Japanese investigators report the alignment of lipid-based myelin figures (Mishima and Morimoto, 1989). Washizu has exploited linearly polarized AC electric fields to align and otherwise manipulate bacterial flagellae and DNA molecules. In one investigation, he oriented bacterial flagellae in a $\sim 5 \cdot 10^5$ V/m electric field at ~ 1 MHz (Washizu, 1990; Washizu et al., 1992a). Measurements of the time required to orient these ~ 10-μm-long by ~ 20-nm-wide flagellae resulted in estimates of their dielectric moment per unit length. Performing somewhat similar experiments with long, helically coiled DNA molecules (length < 20 μm, diameter ~ 2 nm), he observed orientation and uncoiling (Washizu et al., 1993b).

B. *Magnetizable particle applications*

Magnetic particle orientation methods facilitate measurement of magnetic anisotropy in both whole crystals and specimens cut from larger crystals. Either long, rod-shaped or thin, disk-shaped particles can be used for measurement, as

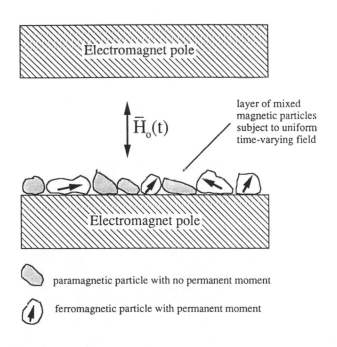

layer of mixed magnetic particles subject to uniform time-varying field

$\bar{H}_o(t)$

paramagnetic particle with no permanent moment

ferromagnetic particle with permanent moment

Fig. 5.9

Magnetic separation method based on the application of a spatially uniform, AC (or pulsating) magnetic field $\bar{H}_0(t)$ to a bed of particles (Andres and O'Reilly, 1992). Ferromagnetic particles having permanent moment $\bar{m}_{perm}$ experience a pulsating torque $\mu_0 \bar{m}_{perm} \times \bar{H}_0(t)$ that effectively fluidizes them. Particles with weak permanent moments long as the orientation of all crystalline axes is known. The typical procedure is to suspend the sample by a thread or fiber in a uniform DC magnetic field and then to measure the torque exerted on the particle. A number of variants of this technique have been exploited over the years, including Weiss's torsion balance method (Bates, 1961).

One new ore beneficiation scheme exploits the alignment torque exerted on particles with a permanent magnetic moment (Andres and O'Reilly, 1992). Conventional MAP separation systems are often ineffective in processing ores containing ferromagnetic and paramagnetic components because the gradient force does not provide sufficient differentiation for commercially practical separation. On the other hand, a far stronger torque can be exerted on ferromagnetic particles than on paramagnetic particles by virtue of the inherent permanent magnetism of the former. Refer to Figure 5.9, which shows a mixture of ferromagnetic and paramagnetic particles resting on the flat face of an AC electromagnet. The ferromagnetic particles, with their inherent permanent dipole moments, experience a pulsating torque $\mu_0 \overline{m}_{\text{perm}} \times \overline{H}_0(t)$ that causes them to dance above the surface in a fluidized state. The paramagnetic particles, having no permanent moment, experience minimal torque and generally adhere to the pole face. This separation concept has achieved good beneficiation of wolframite in tungsten ore.

6

Theory of particle chains

6.1 Introduction to chaining and review of previous work

Closely spaced particles subjected to an electric or magnetic field interact electromechanically through the agency of their induced dipole and higher-order moments. Depending on the relative alignment of the particles, the mutual forces of interaction can be attractive or repulsive. In general, similar particles attract each other when aligned parallel to an applied field and repel each other when in perpendicular alignment. At close spacings, these interaction forces can become quite strong. There are two complementary physical interpretations of such interparticle electromechanics: one approach considers dipole–dipole interactions, while the other focuses upon the distortion of the applied field in the vicinity of each particle. The dipole–dipole interaction model stems from the "action-at-a-distance" physical interpretation of electrodynamics. Figure 6.1a shows two similar particles oriented with their line of centers parallel to the applied field. The induced dipoles clearly attract each other, irrespective of the sign of the Clausius–Mossotti function K (defined in Table A.1 in Appendix A), just as do two permanent magnets when similarly aligned. On the other hand, when the particles are aligned perpendicular, as depicted in Figure 6.1b, they repel each other, again just as two permanent magnets aligned side by side would do.

The other interpretation of field-induced particle interactions focuses on the localized field disturbance caused by dielectric, conductive, or magnetic particles. For $K > 0$, the field is intensified near the poles and reduced near the equator. The reverse occurs when $K < 0$. Thus, the predicted particle interactions are consistent with those derived from consideration of dipole-dipole interactions. An interesting special case, not depicted in Figures 6.1a and b, occurs when $K > 0$ for one particle and $K < 0$ for the other particle. Here, the two particles repel each other when aligned end to end and attract each other in side-by-side orientation. Such behavior, evident from a consideration of the dipole interactions, is also borne out by a field calculation approach (Stoy, 1989a,b).

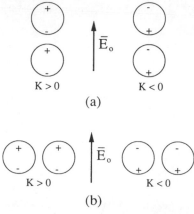

(a)

(b)

Fig. 6.1 Interacting particles with induced dipole moments: (a) parallel (∥) and (b) perpendicular (⊥) alignment with respect to the applied electric field. If K has the same sign for both particles, then the interaction is always attractive for parallel alignment and repulsive for perpendicular alignment.

The attractive forces between similar particles lead to chain formation, a ubiquitous phenomenon with far-reaching implications in particulate science and technology. Though certainly not first to observe electric field–induced chaining, Muth (1927) made detailed observations of the phenomenon using insoluble fat particles suspended in water, calling these formations *Perlschnüre*, or pearl chains. Similar pearl chain structures were observed when suspended biological cells and bacteria were subjected to an alternating electric field (Krasny-Ergen, 1936; Liebesny, 1939). In 1949, it was discovered that particle chain formation in dielectric liquids has a strong influence on the effective viscosity (Winslow, 1949). Füredi and Valentine (1962) reported extensive observations on the formation of particle chains for a large number of nonorganic particles such as polystyrene, ion-exchange resin particles, potato starch, ferric oxide, carbon black, and aluminum powder. These experiments, conducted in insulating oils and (relatively) conductive water at electric field frequencies from DC to 10^7 Hz, revealed that electric field–induced chain formation is often frequency-dependent.

Schwan and co-workers studied pearl chains and proposed a simple theory for predicting both the electric field threshold and the time constant associated with their formation (Saito and Schwan, 1961; Schwan and Sher, 1969). Using a dipole interaction model, they postulated a threshold value of the electric field E_{th} based on a balance of electrostatic energy and thermal energy.

$$E_{th} \approx \frac{1.7 R^{-3/2}}{|K|} \sqrt{\frac{kT}{\varepsilon_2}} \qquad\qquad (6.1)$$

where T is the temperature and $k = 1.38 \cdot 10^{-23}$ J/deg is the Boltzmann constant. This threshold value is inversely proportional to the square root of the particle volume. More recently, AC electric fields have been used to create chains of cells that are then pulsed by a strong DC field to initiate fusion (Zimmermann and Vienken, 1982).

Chains of nonspherical cells exhibit the frequency-dependent orientation effects described in Chapter 5. One example is provided by turkey erythrocytes, which have a flattened ellipsoidal shape with three distinct axes. Below ~7 MHz, these erythrocytes form chains with all cells oriented with their longest axis parallel to the field. Between ~8 MHz and ~22 MHz, this orientation changes so that the intermediate axis is aligned with the field. Above ~26 MHz, all particles align with their shortest axis parallel to the field so that the chain takes on the appearance of a stack of pancakes. In addition, chained cells exhibit electrorotation in the narrow-frequency bands between the different orientational regimes (Zimmermann et al., 1984). The average measured length of pearl chains of melanocytes depends on the AC field frequency and, further, these lengths correlate well to DEP collection spectra obtained by independent experimental means (Mischel et al., 1983).

Field-induced chaining of magnetizable particles has major implications in a number of technologies, from mechanical clutch mechanisms utilizing magnetic slurries or dry powders to magnetic field–stabilized fluid beds to magnetic brush xerographic copier engines. An important early contribution concerning magnetic particle chain behavior was the experimental investigation of Harpavat (1974), who performed simple experiments with model chains to ascertain which link is the weakest in a long chain. He proposed a simple dipole interaction model that achieved qualitative success in predicting the chain behavior. We base the multipolar interaction model developed in this chapter largely on the foundation laid by Harpavat's work.

An important limitation in Chapters 6 and 7 is that only linear multipolar moments are considered; the justification for this restriction is that most of the technologically important particle electromechanics can be adequately described in terms of the strong interactions that occur when particles are aligned with their line of centers parallel to applied fields. Particle interactions in chains oriented perpendicular to applied fields are substantially weaker. The general multipolar interaction model is of far greater importance in modeling the effective dielectric constant of random arrays of clusters or chains (Bedeaux et al., 1987).

Another limitation of both Chapters 6 and 7 is that we restrict attention to spherical models for particles. To the author's knowledge, no model for chain formation of ellipsoids or other nonspherical particles exists at the present time. Chaining of such particles involves a straightforward extension of the results of Chapter 5 coupled with a more general application of the method of effective multipoles.

6.2 Linear chains of conducting spheres

The problem of two closely spaced, *perfectly conducting* spheres in the presence of a uniform electric field is fundamental to chaining because here the particle interactions are strongest. For closely spaced conducting spheres, the poor convergence of standard series solutions presents challenging computational difficulties. Most of the published analytical treatments of the problem fall into one of three categories: (i) noncontacting spheres (Davis, 1964; Arp and Mason, 1977); (ii) spheres in tangential contact (Lebedev and Skal'skaya, 1962; Jeffery and Onishi, 1980); and (iii) intersecting spheres (Jones, 1987; McAllister, 1988). The papers cited above are merely representative and do not constitute a complete set of the published work on problems in this class.

A. Solution using the method of images

Though the *method of images* can be more generally applied, here for convenience we restrict attention to chains of equal-sized spheres. The closely related *method of geometric inversion*, exploited in Section 6.2B, provides a compact, unified treatment of both parallel and perpendicular chains of two unequal spheres. Throughout this chapter, results obtained using either method are compared to available published analytical solutions wherever possible.

It is important to note that the method of images is applicable only to perfectly conducting spheres, that is, $\sigma_2/\sigma_1 \to \infty$. This limit is analogous to $\varepsilon_2/\varepsilon_1 \to \infty$ and $\mu_2/\mu_1 \to \infty$. Therefore, the results obtained in this section have special value because they establish the $K = 1$ limit for the effective moments of chains consisting of dielectric or magnetic particles (c.f., Section 6.3).

Before considering chains, it is instructive first to treat the problem of a single conductive sphere ($N = 1$) immersed in dielectric fluid of permittivity ε_1 and positioned midway between the two source charges $+Q$ and $-Q$, as shown in Figure 6.2a. Image charges $-q$ and $+q$ are induced in the sphere at points along the axis defined by the source charges at distances $-d$ and $+d$, respectively, from the sphere's midpoint. From the theory of images for spherical conductors (Lorrain and Corson, 1970), the values of q and d are

$$q = -QR/D \tag{6.2a}$$

$$d = R^2/D \tag{6.2b}$$

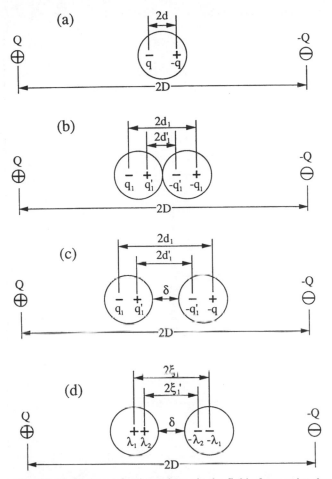

Fig. 6.2 Perfectly conducting spheres in the field of two point charges $+Q$ and $-Q$. (a) Single conductive sphere with image charges. (b) Two touching spheres with primary images. (c) Two noncontacting spheres, showing the primary images. (d) Neutralizing charges and first set of images.

To test these results, use Coulomb's law to write a relationship between the source charge Q and the net effective electric field E experienced at the center of the conductive particle. From superposition,

$$E_0 = Q/2\pi\varepsilon_1 D^2 \tag{6.3}$$

We find from Appendix B that $p^{(1)} = -2qd$. Then, using Equations (6.2a,b) and (6.3), we obtain

$$p^{(1)} = 4\pi\varepsilon_1 R^3 E_0 \tag{6.4}$$

As expected, this expression is identical to the $K \to 1$ limit of Equation (2.12).

A.1 Chain of two contacting particles

Consider $N = 2$ identical perfectly conductive spheres of radius R that are in contact and so oriented that their line of centers includes the charges $+Q$ and $-Q$, as shown in Figure 6.2b. The primary images, $\pm q_1$ and $\pm q_1'$, with their locations, $\pm d_1$ and $\pm d_1'$, are also shown in this figure. Using Equations (6.2a, b), we have

$$q_1 = -RQ/(D - R) \quad \text{and} \quad d_1 = RD/(D - R) \tag{6.5a}$$

$$q_1' = RQ/(D + R) \quad \text{and} \quad d_1' = RD/(D + R) \tag{6.5b}$$

It will be apparent that in fact there must be an infinite number of image charges. Thus, for the ith set of images, we may write (Jones, 1986b)

$$q_i = -RQ/(iD - R) \quad \text{and} \quad d_i = RD/(iD - R) \tag{6.6a}$$

$$q_i' = RQ/(iD + R) \quad \text{and} \quad d_i' = RD/(iD + R) \tag{6.6b}$$

These recursive relations are good for all $i \geq 1$.

As a test of these relations, we calculate the net charge in the left-most sphere $Q_{\text{left}} (= -Q_{\text{right}})$ for the important case of a uniform electric field, imposing $D/R \to \infty$ and $Q \to \infty$ limits in such a way that $E_0 \propto Q/D^2$ remains finite.

$$Q_{\text{left}} = -2Q\left(\frac{R}{D}\right)^2 \sum_{i=1}^{\infty} \frac{1}{i^2} = -(2\pi^3/3)\varepsilon_1 R^2 E_0 \tag{6.7}$$

The summation has been identified as the Riemann-zeta function ($\Sigma i^{-2} \equiv \zeta(2) = \pi^2/6$) and its value substituted into Equation (6.7). This result is the well-known Maxwell charge induced within a conducting sphere lying on a conductive plane surface and subjected to a uniform electric field E_0 (Maxwell, 1954, article 175).

We may calculate the effective dipole moment of the two-particle chain using the above expressions for q_i, d_i, etc., by taking the same $D/R \to \infty$ and $Q \to \infty$ limits. From the definition of the dipole moment in Appendix B, Equation (B.8), we have

$$p_{\text{eff}}^{(1)} = -\sum_{i=1}^{\infty} 2(q_i d_i + q_i' d_i') \tag{6.8a}$$

$$= 8R^3 Q/D^2 \sum_{i=1}^{\infty} \frac{1}{i^3} \tag{6.8b}$$

The summation in Equation (6.8b) is the Riemann-zeta function with an argument value of 3, that is, $\Sigma\, i^{-3} \equiv \zeta(3) = 1.20206$. Therefore

$$p_{\text{eff}}^{(1)} = 16\zeta(3)\pi\varepsilon_1 R^3 E_0 \approx 19.23\,\pi\varepsilon_1 R^3 E_0 \tag{6.9}$$

The important result of Equation (6.9) is that the net moment of two conductive spheres in contact is a factor of $2 \times \zeta(3) \cong 2.4041$ larger than the sum of two identical, noninteracting spheres. We will learn later that this enhancement of the effective dipole moment of the chain is a manifestation of the induced higher-order multipoles.

A.2 Longer chains

The method of images provides a convenient means to calculate the net effective dipole moment for longer linear chains, that is, $N > 2$. Invoking this method for $N = 3$ and 4, we may express the moments as double sums reasonably well suited for numerical reckoning (Jones, 1986b). For longer chains ($N > 4$), the summations become more troublesome, necessitating approximate methods to estimate $p_{\text{eff}}^{(1)}$ (Meyer, 1994). Table 6.1 summarizes calculated results for the effective moments of linear chains consisting of up to $N = 7$ conducting spheres.

For chains of more than two spheres, the particle interactions already evident in Equation (6.9) become even more pronounced. The strength of these interactions is best revealed in terms of the average, normalized dipole moment per particle. This quantity, $p_{\text{eff}}^{(1)}/N = p_{\text{eff}}^{(1)}/p_0 N$, is a very strong function of N. These results along with some confirming experimental data are detailed in Section 6.2F.

A.3 Noncontacting particles

Effective moment calculation for linear chains of noncontacting particles presents the complication of identifying and then imposing the proper electrical constraints on the spheres. We may identify two useful limiting cases: the *isolated particle* and *equipotential* constraints. The isolated particle limit, where each individual particle is constrained to zero net charge, applies if the adjacent particles never make electrical contact. The equipotential case, where the spheres are constrained to be at the same electrostatic potential, describes the physical situation of two conductive particles right after an electrical breakdown has occurred between them.

Consider the two noncontacting, conductive spheres at center-to-center spacing $2R + \delta$ shown in Figure 6.2c. Once again, the primary image charges, q_1 and q_1', and their locations along the system axis, d_1 and d_1', require definition (Jones, 1986b).

Table 6.1 *Effective dipole moments of linear chains of conductive particles calculated using the method of images*

Chain length $(N)^*$	Effective dipole moment of chain $p_{\text{eff}}^{(1)}$	Normalized moment[**] per particle $p_{\text{eff}}^{(1)}/N$	Estimated uncertainty
1	$4\pi\varepsilon_1 R^3 E_0$	1	—
2	$19.23\,\pi\varepsilon_1 R^3 E_0$	2.40	—
3	$48.5\,\pi\varepsilon_1 R^3 E_0$	4.04	<1%
4	$95.2\,\pi\varepsilon_1 R^3 E_0$	5.95	<1%
5	$\sim 161\,\pi\varepsilon_1 R^3 E_0$	~ 8.1	±2%
6	$\sim 250\,\pi\varepsilon_1 R^3 E_0$	~ 10	±5%
7	$\sim 365\,\pi\varepsilon_1 R^3 E_0$	~ 13	±12%

* Moment values for $N = 5, 6$, and 7 are due to Meyer, 1994.
** We use the convenient normalization $p_{\text{eff}} = p_{\text{eff}}/p_0$, where $p_0 \equiv 4\pi\varepsilon_1 R^3 E_0$ is the moment of a single conductive sphere of radius R.

$$q_1 = -RQ/(D - X) \quad \text{and} \quad d_1 = X + R^2/(D - X) \tag{6.10a}$$

$$q_1' = RQ/(D + X) \quad \text{and} \quad d_1' = X - R^2/(D + X) \tag{6.10b}$$

where $X \equiv R + \delta/2$. The recursive relations for the higher-order image charges are

$$q_i = q_{i-1}R/(X + d_{i-1}) \quad \text{and} \quad d_i = X - R^2/(X + d_{i-1}) \tag{6.11a}$$

$$q_i' = q_{i-1}'R/(X + d_{i-1}') \quad \text{and} \quad d_1' = X - R^2/(X + d_{i-1}') \tag{6.11b}$$

This set of images produces a solution that constrains the electrostatic potential of both conductive spheres to zero, corresponding to the equipotential limit.

We distinguish the isolated sphere limit from the equipotential case by a set of neutralization charges that must be added in order to constrain the net charge of each particle to zero. First, we place charges $+q_c$ and $-q_c$ at the centers of the left and right spheres, respectively, and then we add images of these charges to satisfy the constraints imposed by the spherical surfaces. As shown in Figure 6.2d, the first pair of these neutralizing images is $+q_c\lambda_2$ and $-q_c\lambda_2$, placed at distances $-\xi_2$ and $+\xi_2$ from the origin, respectively. Note that $\lambda_1 \equiv 1$ and $\xi_1 \equiv X$. The required recursion relationships are

$$\lambda_i = \lambda_{i-1} R/(X + \xi_{i-1}) \text{ and } \xi_i = X - R^2/(X + \xi_{i-1}) \tag{6.12}$$

To meet the isolated sphere condition, the net charge on each particle is set to zero.

$$\sum_{i=1}^{\infty}(q_i + q_i') + q_c \sum_{i=1}^{\infty} \lambda_i = 0 \tag{6.13}$$

Equation (6.13) is used to determine q_c by numerical computation of the summations.

Using the above results for the image and neutralization charges, we can calculate the effective moment by invoking Equation (B.8) from Appendix B. The expressions for the two limiting cases are:

Isolated sphere case: $\quad p_{\text{eff}}^{(1)} = -\sum_{i=1}^{\infty} 2(q_i d_i + q_i' d_i' + q_c \lambda_i \xi_i) \tag{6.14a}$

Equipotential sphere case: $\quad p_{\text{eff}}^{(1)} = -\sum_{i=1}^{\infty} 2(q_i d_i + q_i' d_i') \tag{6.14b}$

Figure 6.3 contains effective dipole moment calculations for the two cases. For touching spheres, that is, $\delta = 0$, Equations (6.14a) and (6.14b) give the same result for the effective moment. Note how $p_{\text{eff}}^{(1)}$ is a monotonically decreasing function of δ/R for the case of isolated spheres and monotonically increasing with δ/R for the case of equipotential spheres. Figure 6.3 also contains plots of the effective dipole moment for a chain of $N = 3$ equally spaced conductive spheres as a function of δ/R for the isolated and equipotential sphere cases.

A.4 Quadrupolar moments of chains

We may also use the method of images to calculate the effective quadrupolar moment of a particle chain. As a preparation for this exercise, consider a single sphere located midway between two identical positive source charges $+Q$. These charges are spaced a distance $2D$ from each other, as shown in Figure 6.4a, so that $E_0 = 0$ at the center of the sphere. We again use Equations (6.2a,b) to determine the magnitudes of the image charges q and their axial locations $\pm d$; however, to maintain the zero net charge condition for the sphere, we must place a charge of $+2q$ at the origin. The resulting charge distribution is the linear quadrupole shown in Figure B.2. From Equation (B.7) in Appendix B, the effective moment is

$$p_{\text{eff}}^{(2)} = 2qd^2 = -2QR^5/D^3 \tag{6.15}$$

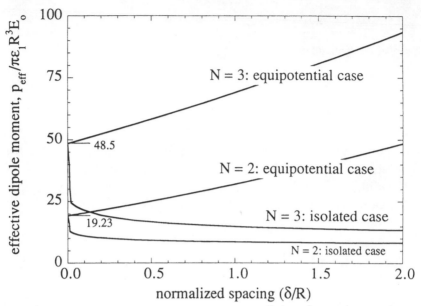

Fig. 6.3 Plots of normalized effective dipole moment $p_{eff}^{(1)}/\pi\varepsilon_1 R^3 E_0$ versus normalized spacing δ/R for linear chains of $N = 2$ and $N = 3$ perfectly conducting spheres oriented parallel to the applied electric field.

We may relate this equation to the first derivative of the electric field by recognizing that $\partial E_0/\partial z = -Q/\pi\varepsilon_1 D^3$ at the center of the sphere.

$$p_{eff}^{(2)} = 2\pi\varepsilon_1 R^5 \frac{\partial E_0}{\partial z} \tag{6.16}$$

which agrees with Equation (2.22) for $n = 2$ in the limit of $\varepsilon_2/\varepsilon_1 \to \infty$.

Extension of this method to calculate the quadrupolar moment of a particle chain is straightforward. Refer to Figure 6.4b, which shows a chain of two identical non-contacting spheres and their associated primary image charges. Equations (6.10a,b) and (6.11a,b) provide magnitudes for all primary image charges, q_i and q_i', and their locations along the axis, d_i and d_i'. The signs of all these charges will be evident from inspection of Figure 6.4b. We again require neutralization charges to achieve the constraint of zero *net* charge for the chain and use Equations (6.12) and (6.13) to evaluate λ_i, ξ_i, and q_c. Using Equation (B.7) from Appendix B, the effective quadrupolar moment is

$$p_{eff}^{(2)} = 2\sum_{i=1}^{\infty}(q_i d_i^2 + q_i' d_i'^2 + q_c\lambda_i\xi_i^2) \tag{6.17}$$

Figure 6.5 plots results for the two-particle chain, normalized by the factor

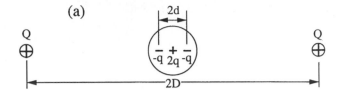

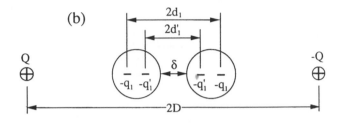

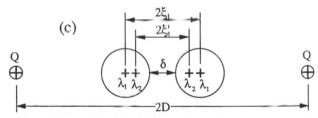

Fig. 6.4 Perfectly conducting spheres in the field of two identical point charges $+Q$ for quadrupole calculation. (a) Single sphere with source and primary image charges. (b) Two noncontacting spheres showing source and primary image charges. (c) Neutralization charges and first set of images.

$\pi \varepsilon_1 R^5 \partial E_0 / \partial z$, versus normalized spacing δ / R. Just as in the case of the dipole moment, interparticle coupling very strongly enhances this quadrupole when $\delta / R \ll 1$.

B. Solution using geometric inversion

The method of geometric inversion involves replacement of the original electrostatics problem with a simpler geometry more amenable to either conventional boundary value techniques or the method of images. The solution to the new problem is then transformed back to the original problem using the rules of geometric inversion. The inversion method realizes its greatest advantage in electrostatics problems where spherical conducting surfaces are present because

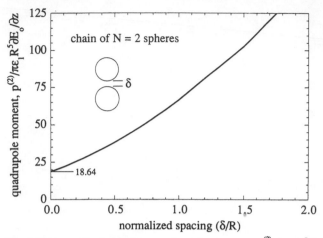

Fig. 6.5 Normalized effective quadrupolar moment $p_{\text{eff}}^{(2)}/\pi\varepsilon_1 R^5 \partial E_0/\partial z$ versus normalized spacing δ/R for linear chain of two identical, perfectly conducting spheres oriented parallel to applied electric field.

spheres invert either to other spheres or to planar surfaces (Smythe, 1968; Section 5.09). The geometric inversion technique particularly lends itself to calculation of $p_{\text{eff}}^{(1)}$ for the case of contacting spheres of unequal radii aligned either parallel or perpendicular to the applied electric field.

B.1 Parallel orientation of identical spheres

To illustrate this method, we revisit the familiar case of two identical conducting spheres of radius R in contact and located midway between the source charges $+Q$ and $-Q$. Figure 6.6a shows the conveniently chosen sphere of inversion having radius $R_{\text{inv}} = 2R$. We seek the induced potential $\Phi(r)$ at some arbitrary position r on the axis. The geometric inversion rule is $rr' = R_{\text{inv}}^2$, where r' (and all other primed variables) refer to the inverted problem. Because both conductive spheres make tangential contact with the inversion sphere, they transform to the pair of grounded parallel planes shown in Figure 6.6b. The source charge pair transforms to a dipole of strength $p' = 16R^3 Q/D^2 = 32\pi\varepsilon_1 R^3 E_0$ located at the origin. In the inverted geometry, this source dipole p' produces an infinite line of dipole images along the axis. The potential $\Phi'(r')$ due to these image dipoles at the point $P'(r' = 4R^2/r)$ on the axis is a sum of dipole potentials.

$$\Phi'(r') = -\sum_{i=1}^{\infty} \frac{p'}{4\pi\varepsilon_1 (4iR - r')^2} + \sum_{i=1}^{\infty} \frac{p'}{4\pi\varepsilon_1 (4iR + r')^2} \tag{6.18}$$

As before, assume that $D/R \to \infty$, while $E_0 \propto Q/D^2$ remains finite. We then obtain

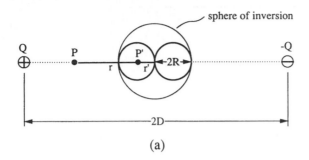

(a)

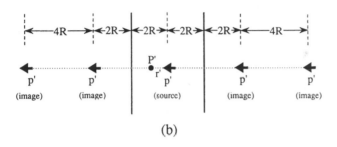

(b)

Fig. 6.6 Method of inversion for chain of two identical, perfectly conducting spheres oriented parallel to the applied electric field. (a) Original problem with two contacting spheres, source charges, and inversion sphere. (b) Inverted problem with source dipole p'

$$\Phi'(r') = \frac{-p'r'}{\pi\varepsilon_1(4R)^3} \sum_{i=1}^{\infty} \frac{1}{i^3} \tag{6.19}$$

Using the inversion rule $\Phi = (2R/r)\Phi'$ and again identifying the Riemann-zeta function, the induced electrostatic potential becomes

$$\Phi(r) = -\frac{4R^3 E_0}{r^2} \zeta(3) \tag{6.20}$$

Comparing this result to Equation (2.7) gives the electrostatic potential at $\theta = 0$ of an electric dipole with moment value

$$p_{\parallel} = 16\zeta(3)\pi\varepsilon_1 R^3 E_0 \tag{6.21}$$

$p_{\parallel}$, signifying the effective dipole moment of a chain of two spheres aligned parallel to the applied electric field E_0, is identical to Equation (6.9).

B.2 Perpendicular orientation of identical spheres

We may adapt the method of inversion to determine the effective moment of a pair of spherical conductors aligned perpendicular to the applied electric field E_0 using

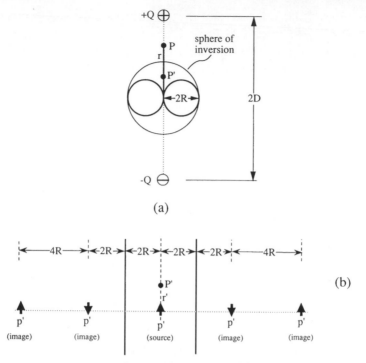

(a)

(b)

Fig. 6.7 Method of inversion for chain of two identical, perfectly conducting spheres oriented perpendicular to the applied electric field. (a) Original problem with two contacting spheres, source charges, and inversion sphere. (b) Inverted problem with source dipole p' and first few image dipoles.

the arrangement shown in Figure 6.7a. Note the equatorial locations now chosen for the source charges $+Q$ and $-Q$. The sphere of inversion is identical to Figure 6.6a. To identify the effective moment, we seek an expression for $\Phi(r)$ at the point r as shown. In the inverted geometry, the source charges transform into a dipole aligned perpendicular to the parallel planes. This dipole induces an infinite set of image dipoles that alternate in sign and are depicted in Figure 6.7b. The potential at $P'(r' = 4R^2/r)$ is expressed in terms of a series due to these image dipoles.

$$\Phi'(r') = \frac{p'}{2\pi\varepsilon_1} \sum_{i=1}^{\infty} \frac{(-1)^i \cos(\pi/2 - \Delta\theta_i)}{16i^2 R^2} \tag{6.22}$$

where the angles are defined by $\Delta\theta_i = \tan^{-1}(r'/4iR)$. If we let $D/R \to \infty$ as before, then we can show that

$$p_\perp = 8\pi\varepsilon_1 R^3 E_0 \sum_{i=1}^{\infty} \frac{(-1)^{i-1}}{i^3} \tag{6.23}$$

$$= 6\zeta(3)\,\pi\varepsilon_1 R^3 E_0 = 7.2123\,\pi\varepsilon_1 R^3 E_0$$

Equation (6.23), which is in agreement with Jeffery and Onishi (1980), reveals that the moment $p_\perp$ is somewhat less than twice the moment of a single sphere, that is, $p_\perp < 2p_0$. This result is a consequence of the mutual depolarization occurring when the spheres are side by side. From a comparison of Equations (6.21) and (6.23), the strong orientational dependence of the effective moment of conducting particle chains is evident.

C. Alignment torque for chain of two identical conducting spheres

With expressions in hand for $p_\parallel$ and $p_\perp$, we may employ superposition to write down the alignment torque exerted upon a rigid chain of two identical conductive spheres of radius R. Consider a rigid two-sphere chain aligned at angle θ with respect to an applied electrostatic field vector $\bar{E}_0$, as shown in Figure 6.8a. The effective moment will be

$$\bar{p}_{\text{eff}} = 8\pi\varepsilon_1 R^3 E_0 \left[\hat{u}_\parallel \left(2\zeta(3) \right)\cos\theta + \hat{u}_\perp \left(3\zeta(3)/4 \right)\sin\theta \right] \tag{6.24}$$

The chain experiences an alignment torque because of this shape-dependent polarization anisotropy of the chain. When Equation (6.24) is plugged into the electrical torque expression, Equation (2.6), the result is

$$T^e = 5\zeta(3)\pi\varepsilon_1 R^3 E_0^2 \sin 2\theta$$
$$\approx 6.01028\,\pi\varepsilon_1 R^3 E_0 \sin 2\theta \tag{6.25}$$

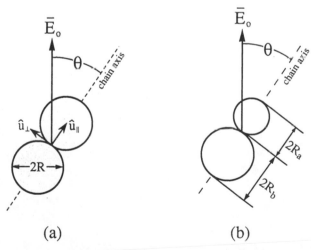

(a)   (b)

Fig. 6.8 Chains of two spheres oriented at angle θ with respect to applied electric field $\bar{E}_0$. (a) Identical spheres of radii R. (b) Two spheres of radii R_a and R_b with $\beta = R_b/R_a > 1$.

This sin 2θ angular dependence is identical to that of Equation (5.19) for the torque on a conductive prolate spheroid. Just like the spheroid, the particle chain is stable at $\theta = 0°$ and unstable at $\theta = 90°$.

D. *Spheres of unequal radii*

The method of inversion adapts readily to the problem of a chain of two unequal conductive spheres. Consider the chain shown in Figure 6.8b, consisting of spheres with radii R_a and R_b, with $\beta = R_b/R_a > 1$. Here, account must be taken of the chain's asymmetry by adding in a monopole to restore the net zero charge condition upon the chain (Jones and Rubin, 1988). To investigate the influence of β upon the effective moment, it is convenient to fix the total volume of the two spheres. To do so, we define $R^3 = R_a^3(1 + \beta^3)/2$, where R represents the radii of two identical spheres in a tangential contact and having the same *total* volume as the chain shown in Figure 6.8b. The dipole moment vector is

$$\bar{P}_{eff} = 8\pi\varepsilon_1 R^3 E_0 \left[\hat{u}_\| \left(\vartheta_\| (\beta) \cos\theta \right) + \hat{u}_\perp \vartheta_\perp (\beta) \sin\theta \right] \tag{6.26}$$

where $\vartheta_\|(\beta)$ and $\vartheta_\perp(\beta)$ are (Jones and Rubin, 1988)

$$\vartheta_\|(\beta) = 2\beta^3 [\Sigma_3 + \Sigma_2^2/4\Sigma_1]/[1 + \beta^3] \tag{6.27a}$$

$$\vartheta_\perp(\beta) = \frac{\beta^3}{[1 + \beta^3]} \sum_{i=1}^{\infty} \left[\frac{2}{[(\beta + 1)i]^3} - \frac{1}{[(\beta + 1)i - 1]^3} - \frac{1}{[(\beta + 1)i - \beta]^3} \right] \tag{6.27b}$$

and the following summations have been defined

$$\Sigma_1(\beta) = \sum_{i=1}^{\infty} \left[\frac{2\beta - i(\beta + 1)^2}{2i(\beta + 1)[(i^2 - i)(\beta + 1)^2 + \beta]} \right] \tag{6.28a}$$

$$\Sigma_2(\beta) = \sum_{i=1}^{\infty} \left[\frac{(2i - 1)(\beta^2 - 1)}{[(i^2 - i)(\beta + 1)^2 + \beta]^2} \right] \tag{6.28b}$$

$$\Sigma_3(\beta) = \sum_{i=1}^{\infty} \left[\frac{2}{[(\beta + 1)i]^3} + \frac{1}{[(\beta + 1)i - 1]^3} + \frac{1}{[(\beta + 1)i - \beta]^3} \right] \tag{6.28c}$$

In the $\beta = 1$ limit, Equations (6.26), (6.27), and (6.28) reduce to Equation (6.24).

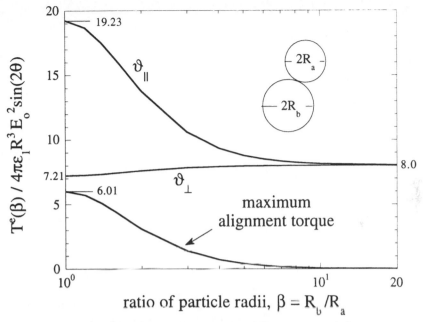

Fig. 6.9 Normalized maximum alignment torque $T^e(\beta)|_{\theta=\pi/4}/4\pi\varepsilon_1 R^3 E_0^2$ versus ratio of radii $\beta = R_b/R_a$. For convenience, the parallel and perpendicular polarization coefficients $\vartheta_\|(\beta)$, and $\vartheta_\perp(\beta)$ defined by Equations (6.27a,b) are also plotted.

An analytical solution to this problem is available (Jeffery and Onishi, 1980), though its results are not easily compared to the above.

The electrical torque $T^e(\beta)$ exerted on the chain shown in Figure 6.8b is

$$T^e(\beta) = 4\pi\varepsilon_1 R^3 E_0^2 \Big(\vartheta_\|(\beta) - \vartheta_\perp(\beta)\Big)\sin 2\theta \qquad (6.29)$$

Figure 6.9 displays the maximum normalized alignment torque, $T^e(\theta = \pi/4)$, as well as $\vartheta_\|(\beta)$ and $\vartheta_\perp(\beta)$, in a plot versus the ratio of the radii β. As expected, the torque goes to zero as $\beta \to \infty$.

E. Intersecting spheres

Another interesting extension of the method of inversion is the case of intersecting spheres, as illustrated in Figure 6.10. Let the angle of intersection be α, the radii of the identical spheres be R_α, and call the center-to-center spacing Ξ. For certain discrete values of the angle α, it is possible to obtain exact expressions for $p_\|$ from a finite number of image charges in the inverted problem (Jones, 1987; McAllister, 1988). Table 6.2 summarizes the values for the parallel effective

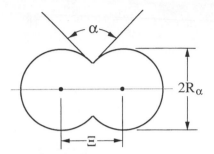

Fig. 6.10 Geometry of two identical intersecting conductive spheres having radii R_α. The degree of intersection is defined by the angle α, or by center-to-center spacing Ξ.

moment ($p_\parallel$) of such intersecting spheres at these discrete angles. Once again, to facilitate meaningful comparison of the results to chains of nonintersecting spheres, we fix the total volume at $8\pi R^3/3$ for all values of α. Therefore, $R_\alpha = R$ represents the case of tangential contact, that is, $\alpha = 0°$.

Figure 6.11 combines results from Table 6.2 for $\Xi < 2R$ with the image theory calculations for noncontacting two-sphere chains ($\Xi > 2R$) in a plot of the effective dipole moment at constant volume. One observes that all discrete points taken from Table 6.2 fall upon a smooth curve linking with the curve derived from image theory using the equipotential constraint. For comparison, the figure also plots the image theory curve for the isolated sphere constraint.

F. Discussion of results for short chains of contacting particles

To summarize the results for linear chains of conductive spheres, it is instructive to plot the calculated values from Table 6.1 for the average effective moment per particle normalized to p_0, $p_{eff}(N)/N$, versus the number of elements in the chain N. For the available values up to $N = 7$, the fitted power-law curve in Figure 6.12 suggests that the moment per particle increases approximately as N^2 for long chains: i.e., $p_{eff}/N \propto N^2$ for $N \gg 1$. This result convincingly demonstrates the strength of the particle–particle interactions for conductive particle chains aligned parallel to the electric field.

The plotted data points in Figure 6.12 are obtained from feedback-controlled DEP levitator measurements performed upon chains of identical metallic particles of diameter ~100 μm (Tombs and Jones, 1991). Refer to Section 3.4D of this book for a description of the DEP levitation method. Agreement of theory and experiment is very good for $N = 2$, but significant scatter emerges for longer chains. We attribute this scatter principally to the presence of surface asperities that strongly influence the spacing δ and the effective dipole moment. A possi-

Table 6.2 *Effective dipole moments of chains consisting of identical spheres intersecting at angle α, as defined in Figure 6.10. All chains have the same total volume of $8\pi R^3/3$*

Intersection angle (α)	Radius (R_α/R)	Norm. spacing (Ξ/R)	Norm. moment ($p_\parallel/\pi\varepsilon_1 R^3 E$)
π^*	$\sqrt[3]{2}$	0	8
$2\pi/3$	1.05827	1.05827	11.85190
$\pi/2$	1.02014	1.44269	14.24126
$2\pi/5$	1.00869	1.63209	15.70716
$\pi/3$	1.00432	1.73954	16.63841
$2\pi/7$	1.00238	1.80623	17.25547
$\pi/4$	1.00141	1.85038	17.68123
$2\pi/9$	1.00089	1.88106	17.98567
$\pi/5$	1.00059	1.90323	18.21018
$\pi/10$	1.00004	1.97545	18.96649
$\pi/25$	1.00000	1.99606	19.18966
0^{**}	1	2	19.23280

* Spheres are completely coincident.
** Spheres are in tangential contact.

Source: Jones, 1987; McAllister, 1988.

ble problem with the $N = 4$ data was that chain length itself was comparable to the electrode spacing. Under these circumstances, one must question the appropriateness of the DEP approximation itself, which assumes that the electric field gradient is weak.

To test the $p_\parallel(\beta)$ component of Equation (6.26), DEP levitation measurements were performed on chains of small, unequally sized metallic spheres. The experimental data plotted in Figure 6.13 correspond well to the theoretical curve. An interesting point about these asymmetric chains is that when their effective moment is measured in a DEP levitator, uncertainty exists about the

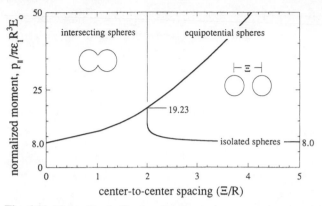

Fig. 6.11 Normalized effective dipole moment $p_\parallel$ versus normalized spacing Ξ/R for particle chains subject to the constant volume constraint. The discrete values obtained for intersecting spheres ($\Xi < 2$) form a curve that joins smoothly with the noncontacting particle curve ($\Xi > 2$) for the equipotential condition.

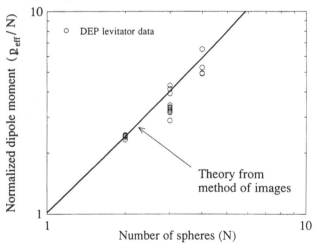

Fig. 6.12 Normalized effective moment per particle p_{eff}/N for chains of identical conducting spheres versus N with experimental data from dielectrophoretic levitation measurements (T. N. Tombs and T. B. Jones, "Effect of moisture on the dielectrophoretic spectra of glass spheres," *IEEE Transactions on Industry Applications*, vol. 29, pp. 281–285, © 1993 IEEE.)

location of the effective midpoint of the chain. Tombs found experimentally that the measured value of $p_\parallel$ for a two-sphere chain is independent of orientation (larger particle on top or bottom) when the geometric center of the chain is used as the reference position in the levitator (Tombs and Jones, 1991). Why the geometric center of the chain should give this result is not immediately evident.

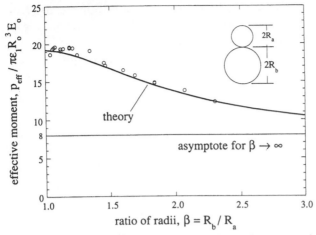

Fig. 6.13 Normalized effective moment for chain of two unequal spheres versus ratio of radii $\beta = R_b/R_a$ with experimental data from dielectrophoretic levitation measurements (T. N. Tombs and T. B. Jones, "Effect of moisture on the dielectrophoretic spectra of glass spheres," *IEEE Transactions on Industry Applications*, vol. 29, pp. 281–285, © 1993 IEEE.)

6.3 Chains of dielectric (and magnetic) spheres

Chain formation is probably of greater technological significance for dielectric (and magnetizable) particles than for conductive particles. Because no image theory exists for spherical dielectric surfaces (Weber, 1965), theoretical treatment of dielectric particle chains requires either a multipolar expansion or solution of a boundary value problem in bispherical or bipolar coordinates. Because of their value in teaching us about the nature of electrostatic interactions between dielectric particles, linear multipoles are emphasized in this book. This choice seems a good one from the standpoint of particle electromechanics because multipolar expansions are directly amenable to interparticle force determination, the subject of Chapter 7.

Analytical and numerical treatments of the important problem of interacting particles have been published for many important cases (Davis, 1969; Godin and Zil'bergleit, 1986; Stoy, 1989a,b; Fowlkes and Robinson, 1988; Chen et al., 1991). Such solutions serve the crucial role of validating the approximate results obtained from multipolar expansions. Where appropriate, we reference some of these analytical solutions in this section.

A dielectric particle in an electric field becomes polarized and the resulting induced moments alter the nearby field. We can always represent these perturbations as a series of multipolar corrections. The simplest example, a dielectric sphere in a uniform electric field, induces a dipolar field distribution from which an effective dipole moment can be extracted. A nonuniform electric field induces higher-order multipolar contributions to the field such as the quadrupole and octupole. Each of these contributions has its own spatial dependence and, consequently, we can identify a set of multipolar moments for any particle in any electric field.

We limit attention here to the simplest case: axisymmetric geometries where linear multipoles adequately represent the induced field contributions. Development of a general model for particle interactions employing effective values for general multipolar moments is certainly possible; however, the most important particle electromechanics can be investigated by restricting attention to axisymmetric problems, the solution of which requires only the linear multipoles.

A. Simple dipole approximation

Consider the problem of two identical dielectric spheres (a and b) of permittivity ε_2 and radius R suspended in a dielectric fluid of permittivity ε_1.[1] The center-to-center spacing is Ξ and the particles are oriented with their line of centers parallel to the uniform applied electric field E_0. First, let us employ the dipole-based approximation introduced by Harpavat (1974) to model the interactions of these two particles. According to this approximation, E_a, the net field at the center of particle a is the sum of the imposed field E_0 and the contribution due to the induced dipole moment p_b of particle b.

$$E_a \approx E_0 + p_b/2\pi\varepsilon_1 \Xi^3 \Big|_{\Xi = 2R} \tag{6.30}$$

Here, symmetry dictates that $p_a = p_b$. From Equation (2.12), we also know that

$$p_a = 4\pi\varepsilon_1 K R^3 E_a \tag{6.31}$$

where $K(\varepsilon_2,\varepsilon_1)$ is the Clausius–Mossotti function defined by Equation (2.13). Equations (6.30) and (6.31) combine to give an expression relating the dipole moment to the imposed electric field.

$$p_a = p_b \approx \frac{K}{1 - K/4} p_0 \tag{6.32}$$

where $p_0 \equiv 4\pi\varepsilon_1 R^3 E_0$, the dipole moment of an isolated perfectly conducting sphere, serves as a convenient normalization. The field interaction has the effect

[1] In this section, we express all results in terms of electric fields and dielectric permittivities. By exploiting the analogy between electric and magnetic systems (c.f., Appendix A), equivalent results for magnetizable particles are obtained readily.

of enhancing the moment as long as $K > 0$. The normalized effective moment of the chain is

$$p_\| = \frac{p_a + p_b}{p_0} \approx \frac{2K}{1 - K/4} \tag{6.33}$$

Employing similar methods, we may find an expression for the effective moment of a chain of two dielectric particles aligned perpendicular to the applied field.

$$p_\perp = \frac{p_a + p_b}{p_0} \approx \frac{2K}{1 + K/8} \tag{6.34}$$

In contrast to $p_\|$ for a chain aligned parallel to the field, now the particle interaction diminishes the net effective moment by a mutual depolarization effect when $K > 0$.

We can easily apply the dipole approximation to longer chains of particles. It is necessary only to write expressions for the electric field at the center of each particle in terms of the various unknown induced moments and then to solve for them. Using this approach, the parallel and perpendicular effective moments of a chain of $N = 3$ identical dielectric spheres are (Jones et al., 1989)

$$p_\| \approx \frac{3K + 31K^2/32}{1 - K/32 - K^2/8} \tag{6.35a}$$

$$p_\perp \approx \frac{3K - 31K^2/64}{1 + K/64 - K^2/32} \tag{6.35b}$$

We may insert either of the above sets of expressions for $p_\|$ and $p_\perp$ into Equation (5.11) to estimate the alignment torque.

The dipole approximation used above ignores all higher-order multipoles so that stringent limits are imposed upon accuracy. By including quadrupolar terms, we can at least estimate the contribution of the higher-order moments to $p_\|$ and $p_\perp$. We rework the problem, replacing Equation (6.30) by a new expression for E_a that includes the field contribution due to the quadrupolar moment $p_b^{(2)}$ induced in particle b.

$$E_a \approx E_0 + \left. p_b^{(1)}/2\pi\varepsilon_1 \Xi^3 \right|_{\Xi = 2R} + \left. 3p_b^{(2)}/4\pi\varepsilon_1 \Xi^4 \right|_{\Xi = 2R} \tag{6.36}$$

The quadrupolar moment is proportional to the derivative of E_a from Equation (2.22).

$$p_a^{(2)} = 4\pi\varepsilon_1 K^{(2)} R^5 \partial E_a/\partial z \tag{6.37}$$

Correct to quadrupolar terms, this derivative is

$$\partial E_a/\partial z \approx -3 p_b^{(1)}/2\pi\varepsilon_1 \Xi^4\Big|_{\Xi = 2R} -3 p_b^{(2)}/\pi\varepsilon_1 \Xi^5\Big|_{\Xi = 2R} \tag{6.38}$$

Combining these equations, we obtain

$$p_\| \approx \frac{2K}{1 - \dfrac{K^{(1)}}{4} - \dfrac{9K^{(1)}K^{(2)}/16}{8 - 3K^{(2)}}} \tag{6.39}$$

where $K^{(1)} = K$ and $K^{(2)}$ is defined by Equation (2.23). Comparing Equation (6.39) to (6.33) shows that, when $\varepsilon_2 > \varepsilon_1$, the quadrupolar term further increases the net moment per particle. In fact, for particle chains in the parallel orientation, all higher-order induced moments have the same effect, that is, they enhance the induced effective dipole moment of each particle in the chain.

Calculation of $p_\perp$ is unaffected by the linear quadrupole because the field contribution due to this moment is zero in the equatorial plane. Therefore, Equation (6.34) may be used to estimate the moment with the particle in the perpendicular orientation $p_\perp$. One might question the accuracy of using only linear multipoles in the case of particle chains aligned perpendicular to the applied field; however, the neglected general multipoles do not significantly influence the net effective dipole moment per particle for layers aligned perpendicular to the field (Miller and Jones, 1988).

B. *Accuracy considerations*

We may demonstrate the failure of the dipole approximation as $K \to 1$ by comparing its results to values obtained for conductive particle chains using the method of images. Such a comparison is permissible because of the analogy between the limits $\sigma_2/\sigma_1 \to \infty$ and $\varepsilon_2/\varepsilon_1 \to \infty$. Table 6.3 contains (i) the dipole approximation results for the moments of each particle in linear chains of up to $N = 6$ identical spheres for $K = 1$, and (ii) values for the moment per particle $p_\|/N$ obtained from the method of images. As expected, particles in the middle of the chain are most strongly polarized. Comparing $p_\|/N$ from the dipole approximation and the method of images shows that the inaccuracy of the dipole approximation rapidly worsens as chain length increases.

Figure 6.14, plotting the net moment normalized by the convenient factor Kp_0 for chains of up to $N = 6$ identical particles versus K, provides a more

Table 6.3 *Values of normalized moments of each particle, $p_1, p_2, \ldots, pN$, in linear chains of N identical particles with K = 1, plus the net ($p_\parallel$) and average ($p_\parallel/N$) moments per particle*

	Results obtained using dipole approximation ($K = 1$)								Image theory
N	p_1	p_2	p_3	p_4	p_5	p_6	$p_\parallel$	$p_\parallel/N$	$p_\parallel/N$
1	1						1	1	1
2	1.33	1.33					2.67	1.33	2.40
3	1.48	1.70	1.48				4.67	1.56	4.04
4	1.55	1.92	1.92	1.55			6.94	1.73	5.95
5	1.59	2.00	2.10	2.00	1.59		9.28	1.86	~8.1
6	1.61	2.04	2.19	2.19	2.04	1.61	11.68	1.95	~10

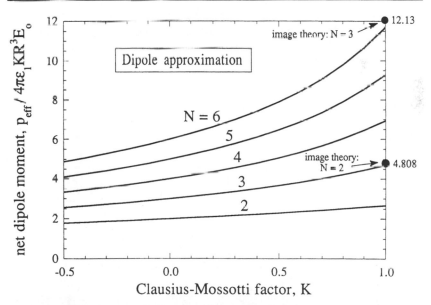

Fig. 6.14 Normalized effective moment calculated using the dipole approximation and plotted versus K for chains of N = 2, 3, 4, 5, and 6 identical dielectric spheres. Image theory results for N = 2 and 3 are plotted along K = 1 axis.

instructive display of how the dipole approximation fails. For reference, values of $p_\parallel$ for $N = 2$ and 3 obtained from the method of images for conductive spheres are plotted as solid circles at $K = 1$ in this figure.

The dependence of $p_\parallel$ upon K for chains of contacting dielectric spheres reveals much about how the higher-order multipoles influence the particle inter-actions. Figure 6.15 plots $p_{\parallel}$, again normalized by Kp_0, versus K for a chain of two identical dielectric spheres obtained using the dipole model of Equation (6.33), the dipole + quadrupole model of Equation (6.39), and a higher-order calculation using 100 terms. Beyond $K \approx 0.5$, the quadrupolar correction does not significantly improve accuracy. It is clear that even 100 multipolar terms becomes inadequate for $K \geq 0.95$. A general methodology for linear multipolar expansions is described in the next section.

C. General expansion of linear multipoles

For improved accuracy in modeling interactions between the particles in a linear chain, we must replace the simple dipole approximation of Section 6.3A with the more general multipolar expansion. Series expressions must be truncated but convergence problems impose limits on accuracy, especially for closely spaced particles when $K \cong 1$. Here, we confine attention to two easily identified limits for particle chains: the chain of two identical spheres ($N = 2$) and the infinite chain ($N \to \infty$). We assume the chains to consist of identical dielectric spheres of permittivity ε_2 and radius R immersed in a fluid of permittivity ε_1.

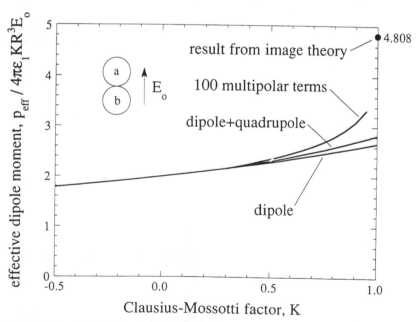

Fig. 6.15 Normalized effective moment of $N = 2$ identical spheres versus K using: the dipole approximation, Equation (6.33); the dipole + quadrupole approximation, Equation (6.39); a linear multipolar expansion using 100 terms (Jones et al., 1989).

For two spheres a and b at center-to-center spacing Ξ aligned parallel to an imposed uniform electric field E_0, all linear multipoles (dipole, quadrupole, octupole, etc.) will be present. Thus, it is necessary to write expressions for all the moments, $p_a^{(i)}$ and $p_b^{(i)}$ with $i = 1, 2, 3, \ldots$, in terms of the axial electric field E_z and its derivatives $\partial E_z / \partial z$, $\partial^2 E_z / \partial z^2$, etc. We may write the following for the field and its axial derivatives at the center of the particle called a.

$$E_a = E_0 + \sum_{i=1}^{\infty} \frac{(i+1)p_b^{(i)}}{4\pi\varepsilon_1 \Xi^{i+2}} \tag{6.40a}$$

$$\frac{\partial^n E_a}{\partial z^n} = \sum_{i=1}^{\infty} \frac{(-1)^n (i+1)(i+2)\ldots(i+1+n)p_b^{(i)}}{4\pi\varepsilon_1 \Xi^{i+2+n}}, n \geq 1 \tag{6.40b}$$

Note that symmetry demands $p_b^{(i)} = (-1)^{i-1} p_a^{(i)}$. Equations (6.40a,b) are in fact merely generalizations of Equations (6.36) and (6.38), written to account for the higher-order induced multipoles. We also need to adapt Equation (2.22), which relates the electric field and its derivatives to the linear multipolar moments.

$$p_a^{(n)} = \frac{4\pi\varepsilon_1 K^{(n)} R^{2n+1}}{(n-1)!} \frac{\partial^{n-1} E_a}{\partial z^{n-1}} \tag{6.41}$$

We obtained the upper curve in Figure 6.15 by truncating the summation in Equations (6.40a,b) at 100 terms and then solving for the net dipole moment (Jones et al., 1989). Even with these 100 terms, the accuracy becomes quite poor for $K \geq 0.95$, which is equivalent to $\varepsilon_2 / \varepsilon_1 \geq 50$.

An especially informative presentation of effective moment calculations for chains of $N = 2$ identical dielectric spheres results by plotting the net moment $p_{\parallel}$ versus $\log_{10}(\delta/R)$ for different values of $\varepsilon_2 / \varepsilon_1$. See Figure 6.16. At spacings greater than a few tenths of a radius, that is, $\delta/R \geq {\sim}0.2$, the mutual interactions are rather insignificant. On the other hand, at closer spacings, they become quite strong and highly dependent on the ratio $\varepsilon_2 / \varepsilon_1$.

We may readily adapt the multipolar expansion method for an infinite chain of identical spherical particles. The key is to recognize that, due to symmetry, only the odd multipoles are induced. We follow the same procedure used above to write an equation for the net electric field E_z and its axial derivatives measured at the center of any particle.

$$E_z = E_0 + \sum_{\substack{i=1 \\ \text{(odd only)}}}^{\infty} \frac{(i+1)\zeta(i+2)p^{(i)}}{2\pi\varepsilon_1 \Xi^{i+2}} \tag{6.42a}$$

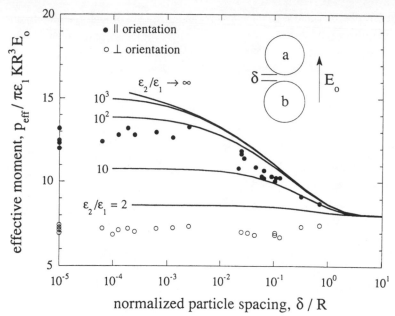

Fig. 6.16 Normalized effective moment of $N = 2$ sphere chains versus normalized spacing δ/R for $\varepsilon_2/\varepsilon_1$ (or μ_2/μ_1) = 2, 10, 100, 1000, and ∞. Data are from vibrating sample magnetometer measurements performed upon 1.59-mm-diameter chrome steel balls (Jones and Saha, 1990).

$$\frac{\partial^n E_z}{\partial z^n} = \sum_{\substack{i=1 \\ \text{(odd only)}}}^{\infty} \frac{(i+1)(i+2)\ldots(i+1+n)\zeta(i+2)p^{(i)}}{2\pi\varepsilon_1 \Xi^{i+2+n}}, n \geq 1 \qquad (6.42\text{b})$$

Equation (6.41), which relates the individual moments to axial derivatives of the field, remains applicable here. Truncation of the infinite series in Equations (6.42a,b) results in a finite set of equations for a finite number of the odd moment terms. In Figure 6.17 we plot versus K normalized dipole moment values *per particle* for contacting spheres ($\Xi = 2$) obtained in this way. Neither the dipole nor the dipole + octupole approximations fare very well for $K \geq 0.5$ when the spheres are in contact. On the other hand, the curve resulting from a calculation using the multipolar moments $p^{(1)}, p^{(3)}, p^{(5)}, \ldots p^{(55)}$ is accurate to within $\pm 1\%$ for $-0.5 \leq K \leq 0.95$ (Jones et al., 1989).

D. *Experimental measurements on chains*

In Figure 6.12, the excellent agreement between the image theory and the DEP measurements for $N = 2$ may be a bit misleading because of the unknown nature

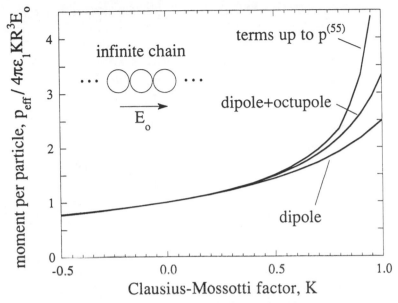

Fig. 6.17 Normalized effective dipole moment per particle for an infinite chain of identical spheres versus K using: the dipole approximation; the dipole + quadrupole approximation; a linear multipolar expansion based on the odd-numbered multipolar terms from $n = 1$ to $n = 55$ (Jones et al., 1989).

of the contact between the conductive particles. According to static theory, two conductive spheres in a uniform DC electric field should not even form a chain but instead should be pulled apart (Lebedev and Skal'skaya, 1962). This prediction is borne out when chains constructed and levitated using an AC electric field are suddenly subjected to a DC electric field (Margolis, 1988). Still, the question remains: why does the chain remain intact in an AC field? The answer is that the dynamic viscosity η_1 of the suspending fluid restricts the rate at which the particles can separate in response to an electric field. If the period of the AC electric field $2\pi/\omega$ in the levitator is much shorter than the electromechanical time constant $\tau_m = \eta_1/4\pi\varepsilon_1 E_0^2$, then field polarity reversals are simply too rapid for any particle response (Arp et al., 1980). In effect, the particles are trapped in a mode where they remain very close together, possibly bouncing off each other at high speed.

For two closely spaced conductive particles in an electric field E_0, the intensified field in the gap between them will most likely exceed the breakdown strength of even good dielectric liquids. An approximate expression for this intensification factor is (Arp and Mason, 1977)

$$\frac{E_{gap}}{E_0} = \frac{-2\pi^2/3}{\frac{\delta}{R}\ln\left(\frac{\delta/R}{12.69}\right)} \tag{6.43}$$

where E_{gap} is the electric field in the gap δ between two identical spheres of radius R. Equation (6.43) is accurate for small spacings, $\delta \ll R$. If E_{gap} exceeds the breakdown strength of the suspending fluid, then the two particles will be driven to the same potential and the effective moment of the chain will be determined by the upper curve in Figure 6.3. In a strong electric field, breakdown should make the effective dipole moment of a chain quite insensitive to the particle spacing. It remains a puzzle why this plausible argument seems not to carry over to longer chains; most of the $p_{\parallel}$ data in Figure 6.12 for 3 and 4 sphere chains fall *below* the prediction of the theory.

Available data for chains of dielectric particles obtained from DEP levitation are limited to variable-frequency investigations of spherical glass beads suspended in silicone oils (Tombs and Jones, 1992). Such chains exhibit distinct Maxwell–Wagner relaxation frequencies reflecting the spectra of the constituent particles. Figures 6.18a and b plot spectra for chains consisting of particles selected to have, respectively, either widely separated or virtually identical Maxwell–Wagner relaxation frequencies. As one might expect, the spectrum of the chain of dissimilar particles evinces the dispersion of each individual particle, while the spectrum of the chain of identical particles has a single relaxation.

Yarmchuk and Janak (1982) performed experiments to investigate the importance of higher-order multipolar moments in controlling the particle interactions between ferromagnetic (steel) balls assembled into linear chains and exposed to low-frequency AC magnetic fields. They studied chains of as many as $N = 10$ identical 0.635-cm-diameter spheres, using plastic shims to fix the interparticle spacings from 15 to ~1600 μm. To measure the magnetization at different locations in chains, they used a one-turn pickup coil wrapped around the equator of a test particle. See Figure 6.19, which shows the experimental arrangements. Experimental data for a chain of $N = 7$ identical spheres correlated reasonably well to numerical calculations obtained from a 30-term truncation of a linear multipolar theory (similar to Section 6.3C), that is, $p^{(1)}$ to $p^{(30)}$, using $\mu_2/\mu_1 \approx 66$ (± 25). The data obtained in these measurements displayed no evidence of nonlinearity, possibly because of a limitation on the maximum AC field achievable with their solenoid.

Using a vibrating sample magnetometer (VSM), Jones and Saha (1990) measured the effective moments of chains of up to $N = 4$ identical chrome steel balls in both parallel and perpendicular alignments. These experiments facilitated the first investigation of the effect of magnetic nonlinearity on chain interactions. Figures 6.20a and b show the measured specific magnetization values for parallel and perpendicular chains of various lengths, respectively. For reference, the magnetization curves for single balls are provided in each figure. The initial slopes of these curves reflect the enhancement or reduction of the normalized

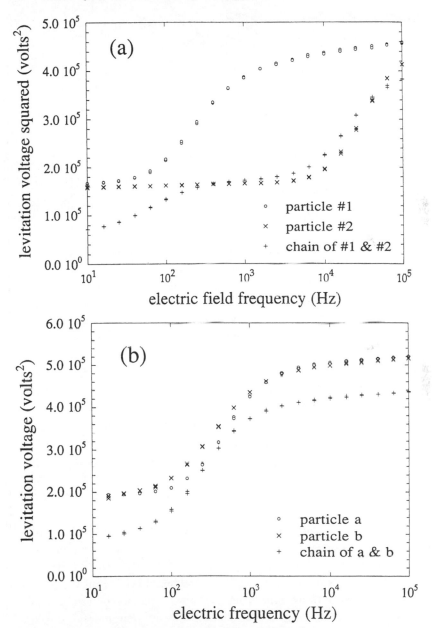

Fig. 6.18 Individual DEP spectra for pairs of 50-μm glass spheres and for chains consisting of these particles. Note that the relaxation frequency of each constituent is reflected in the chain spectrum. (from T. N. Tombs and T. B. Jones, "Effect of moisture on the dielectrophoretic spectra of glass spheres," *IEEE Transactions on Industry Applications,* vol. 29, pp. 281–285, © 1993 IEEE.) (a) Particles with dissimilar relaxation frequencies. (b) Particles with virtually identical relaxation frequencies.

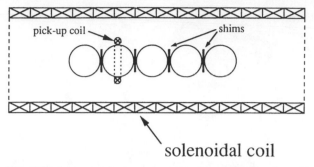

solenoidal coil

Fig. 6.19 Apparatus for measurement of the magnetization of individual spherical particles in linear chains (Yarmchuk and Janak, 1982). The particle with the one-turn pickup coil can be placed at any position in the chain and is used to measure the effective induced moment of particles at different locations in the chain.

moment per particle for the parallel ($p_{\parallel}/N$) and perpendicular ($p_{\perp}/N$) orientations, respectively. Likewise, the shapes of the curves indicate how chain length and orientation influence the nonlinear magnetic responses of the chrome steel balls comprising the chains. Longer chains oriented parallel to the magnetic field approach saturation more rapidly, an expected result, given that the magnetic field is strongly intensified at the poles and reduced at the equator of a single particle. Field intensification at the poles means that regions close to contact points between adjacent particles saturate before the rest of the particle volume. On the other hand, neither chain length nor orientation influences the asymptotic effective specific magnetization.

Representative effective moment data from VSM measurements for two-sphere chains are plotted versus normalized spacing δ/R in Figure 6.16. The data at the larger spacings were obtained using chains assembled and glued into a sample holder, while the data for $10^{-4} < \delta/R < 10^{-2}$ were obtained using balls upon which thin gold (Au) layers had been vacuum-deposited. All data plotted along the left border (at $\delta/R = 10^{-5}$) represent uncoated particles held in mechanical contact. The theoretical curves superimposed on this figure are from multipolar calculations at several values of the relative permeability $\mu_r \equiv \mu_2/\mu_1$, with $\mu_1 = \mu_0$ (Fowlkes and Robinson, 1988). The best fit is achieved for $30 < \mu_r < 50$, which corresponds reasonably well to accepted initial permeability values of ferromagnetic steels. Despite the chain formation, saturation in ferromagnetic particles does not occur until the external field is fairly strong, a result suggesting that interparticle field intensification is very strongly dependent upon μ_r.[2]

---

[2] Equation (6.43) cannot be adapted to estimate H_{gap}, the magnetic field intensity in the gap, because μ_2/μ_1 is not large enough for any ferromagnetic particle.

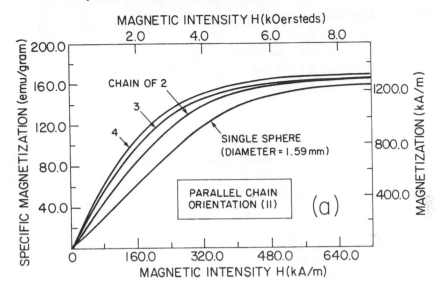

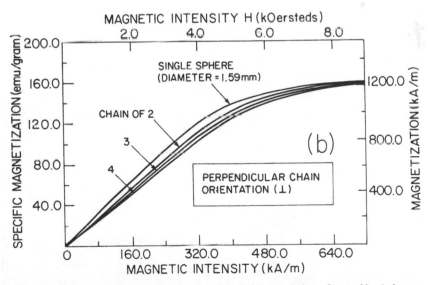

Fig. 6.20 Effective specific magnetization for chains consisting of up to $N = 4$ chrome steel balls (1.59 mm in diameter) measured using a vibrating sample magnetometer. Comparison of the chain magnetization curves to the curve for a single ball shows that chaining influences the initial magnetization per particle but not the asymptotic saturation value (Jones and Saha, 1990). (a) Parallel chains. (b) Perpendicular chains.

The effect of ferromagnetic saturation upon the interparticle forces between particles is discussed in Chapter 7.

For comparison, data for perpendicular chains of ferromagnetic steel balls also have been plotted in Figure 6.16. The rather modest demagnetization effect predicted by Equation (6.34) is evident.

6.4 Frequency-dependent orientation of chains

In Chapter 5, we learned that particles possessing any kind of polarization anisotropy will experience a torque that tends to align the particle with some preferred axis parallel to the applied electric field. When such particles have either ohmic loss or any type of dielectric dispersion, they can exhibit a frequency-dependent orientational effect called an orientational spectrum. Linear particle chains are perfectly good examples of objects with shape-dependent anisotropy and they indeed exhibit turnover phenomena (Füredi and Ohad, 1964). Such turnovers may have important implications for electrorheological (ER) fluids, which are known to exhibit frequency-dependent behavior (Klass and Martinek, 1967).

To develop a model for the frequency-dependent orientational behavior of dielectric particle chains, we take the same approach used in Section 5.3. It is only necessary to generalize the expressions for $p_\parallel$ and $p_\perp$ from Section 6.3A to accommodate particle dispersion by replacing static permittivities with their complex, frequency-dependent equivalents, that is, $\varepsilon \to \underline{\varepsilon}$. To simplify the analysis, only dipole interactions are considered. Consider a *rigid* chain of two identical spheres with complex permittivity $\underline{\varepsilon}_2$ in a suspending fluid of complex permittivity $\underline{\varepsilon}_1$, similar to the chain depicted in Figure 6.8a. The axis of the chain makes an angle θ with respect to linearly polarized electric field vector

$$\bar{E}_0(t) = \mathrm{Re}[(\bar{E}_\parallel + \bar{E}_\perp)\exp(j\omega t)] \tag{6.44}$$

Invoking the superposition principle, we express the effective dipole moment of the chain using Equations (6.33) and (6.34). The result is

$$\bar{\underline{p}}_{\mathrm{eff}} = 4\pi\varepsilon_1 R^3 [\underline{K}_\parallel \bar{E}_\parallel + \underline{K}_\perp \bar{E}_\perp] \tag{6.45}$$

with the following definitions:

$$\underline{K}_\parallel \equiv \frac{2\underline{K}}{1 - \underline{K}/4} \quad \text{and} \quad \underline{K}_\perp \equiv \frac{2\underline{K}}{1 + \underline{K}/8} \tag{6.46}$$

From Equation (4.3a), the time-average torque of electrical origin is

$$\langle \overline{T}^e \rangle = 2\pi\varepsilon_1 R^3 \text{Re}[\underline{K}_\| - \underline{K}_\perp](\overline{E}_\| \times \overline{E}_\perp) \tag{6.47}$$

It is convenient to write the frequency-dependent term in Equation (6.47) as a fractional polynomial in $\underline{K}$.

$$\text{Re}[\underline{K}_\| - \underline{K}_\perp] = \text{Re}\left[\frac{24\underline{K}^2}{(4 - \underline{K})(8 + \underline{K})}\right] \tag{6.48}$$

The zeros of this expression determine the frequencies at which a rigid chain will turn over. The existence of such zeros for real frequencies is not guaranteed, but if they do exist, they always occur in pairs. Just as with ellipsoidal particles, parallel orientation is always favored at the lowest and highest frequencies; if turnovers do exist, the chain first changes from parallel to perpendicular orientation and then back to parallel as frequency is increased. Figures 6.21a, b, c, and d map the regions of stable orientation (parallel and perpendicular) versus electric field frequency and $\varepsilon_2/\varepsilon_1$ for various values of σ_2/σ_1.

The orientational spectra predicted by this simple dipole theory are surprisingly insensitive to either higher-order moments or chain length. Perhaps the most important result is that the frequency band for the perpendicular orientation of chains generally brackets the peak of the electrorotational spectrum (Jones, 1990). Relating these predictions to the experimental observations with cell chains is difficult because the cells are not rigidly connected. Even if a chain does start to turn over due to the orientational instability, the cells are still free to separate and/or rotate independently (Holzapfel et al., 1982), thus greatly complicating the nature of the system under investigation.

6.5 Heterogeneous mixtures containing particle chains

Because of the strong interactions among closely spaced particles with high relative dielectric constant, particle chaining in a solid/liquid suspension has a dramatic effect on the effective dielectric constant of the heterogeneous medium. We can understand this effect by recalling, from Table 6.1 and Figure 6.12, that the effective moment *per particle* is strongly dependent on chain length N. This effect is analogous to the well-known influence of particle shape upon effective dielectric constant and electrical conductivity for mixtures (Sillars, 1937). We introduce below a simple model for heterogeneous mixtures of agglomerated particles to illustrate how chaining influences the effective bulk dielectric permittivity. Electric field–induced orientation of suspended submicron particles and agglomerates also influences certain important optical properties of colloidal media (Stoylov, 1991).

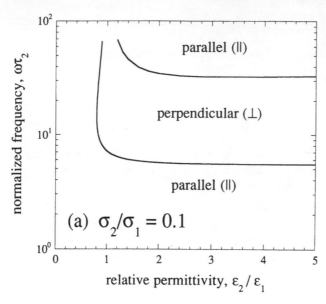

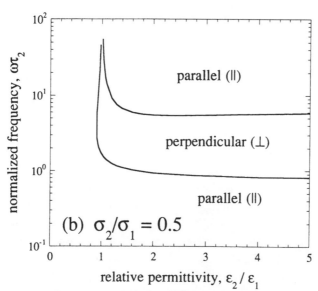

Fig. 6.21 Parallel (∥) and perpendicular (⊥) orientation of rigid two-sphere chains as a function of normalized frequency $\omega\varepsilon_2/\sigma_2$ and relative permittivity $\varepsilon_2/\varepsilon_1$ for various values of the relative conductivity. (a) $\sigma_2/\sigma_1 = 0.1$; (b) $\sigma_2/\sigma_1 = 0.5$; (c) $\sigma_2/\sigma_1 = 2.0$; (d) $\sigma_2/\sigma_1 = 10.0$ (Jones, 1990). Parts (c) and (d) are on facing page.

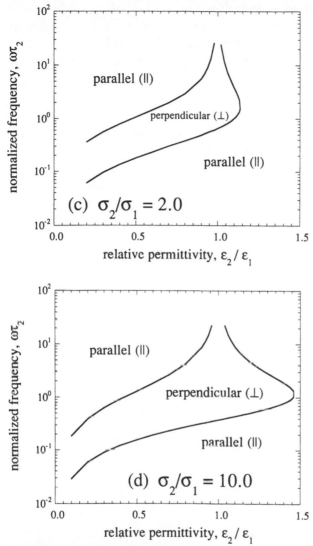

Fig. 6.21 Caption on facing page.

A. Mixture theory

The basic theory of mixtures is attributed to Maxwell, who considered the important case of a heterogeneous conducting medium consisting of a dilute dispersal of spherical particles suspended in a uniform continuous phase. The fundamental assumption of this model is that interactions among particles are completely ignored. It is a straightforward matter to adapt Maxwell's original formulation of the mixture problem to the case of dispersed dielectric particles by using the analogy between conductors and dielectrics described in Appendix A. Consider a spherical ensemble of spherical dielectric particles of permittivity ε_2 in a fluid medium of permittivity ε_1 and subjected to a uniform electric field E_0 as shown in Figure 6.22a. Though unnecessary to do so, we assume here that all the dispersed spheres have the same radius R, which is much smaller than R_0, the radius of the ensemble, that is, $R \ll R_0$. The objective is to obtain an expression for the effective permittivity ε_{eff} of an equivalent sphere of radius R_0, as shown in Figure 6.22b, which produces the same induced dipole moment as the original heterogeneous ensemble.

The approach utilized to obtain ε_{eff} relies on the same effective moment concept favored throughout this book. If the effective moment of each non-interacting particle is p_{particle} and there are a total of N_p particles dispersed within the spherical volume of radius R, then, by superposition

$$p_{\text{eff}} = N_p p_{\text{particle}} \tag{6.49}$$

The effective dipole moment of the equivalent sphere is

$$p_{\text{eff}} = 4\pi R_0^3 \varepsilon_1 \left[\frac{\varepsilon_{\text{eff}} - \varepsilon_1}{\varepsilon_{\text{eff}} + 2\varepsilon_1} \right] E_0 \tag{6.50}$$

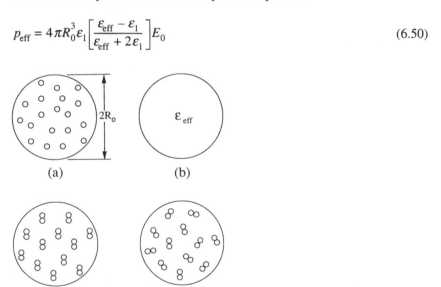

(a)   (b)

(c)   (d)

Fig. 6.22 Spherical ensembles of radius R_0 for calculation of effective dielectric permittivity. (a) Sphere of radius R_0 containing dispersal of smaller spheres of radius R and permittivity ε_2 in continuous medium of permittivity ε_1. (b) Equivalent sphere of radius R_0 and permittivity ε_{eff}. (c) Ensemble of oriented chains. (d) Ensemble of randomly oriented two-sphere chains.

If we now employ the expression for the effective moment of a single particle, that is, $p_{particle} = 4\pi R^3 \varepsilon_1 K(\varepsilon_2, \varepsilon_1) E_0$, the result is Maxwell's mixture formula (Maxwell, 1954, article 314).

$$\frac{\varepsilon_{eff} - \varepsilon_1}{\varepsilon_{eff} + 2\varepsilon_1} = \varphi_1 \left[\frac{\varepsilon_2 - \varepsilon_1}{\varepsilon_2 + 2\varepsilon_1} \right] \tag{6.51}$$

where $\varphi_1 \equiv N_p(R/R_0)^3$ is the volume fraction of the spheres in the mixture.[3] We can solve Equation (6.51) explicitly for ε_{eff} to obtain the common form of the mixture formula.

$$\varepsilon_{eff}/\varepsilon_1 = \frac{1 + 2\varphi_1 K(\varepsilon_2, \varepsilon_1)}{1 - \varphi_1 K(\varepsilon_2, \varepsilon_1)} \tag{6.52}$$

We can generalize Maxwell's approach without difficulty for dispersions of nonspherical particles. For example, consider an ensemble of identical, uniformly aligned ellipsoids with semi-major axes a, b, and c. To obtain expressions for the effective permittivity of the heterogeneous medium along the three orthogonal axes, we define a geometrically similar heterogeneous ellipsoid with semi-major axes A, B, and C, where $A:B:C = a:b:c$. From Equations (5.2), (5.3), and (6.49), the result is

$$\frac{\varepsilon_{eff,x} - \varepsilon_1}{1 - \left(\dfrac{\varepsilon_{eff,x} - \varepsilon_1}{\varepsilon_1} \right) L_x} = \varphi_1 \left[\frac{\varepsilon_2 - \varepsilon_1}{1 - \left(\dfrac{\varepsilon_2 - \varepsilon_1}{\varepsilon_1} \right) L_x} \right] \tag{6.53}$$

with similar expressions for the y and z components. These results may be manipulated to obtain explicit expressions for $\varepsilon_{eff,\,x}$, etc. As expected, a heterogeneous medium of dispersed ellipsoids exhibits dielectric anisotropy.

B. Suspensions of chains

The key to finding an expression for the effective permittivity of a heterogeneous mixture of suspended particle chains is to find the appropriate expression for $p_{particle}$ and then to use it in Equation (6.49). Again, imagine a sphere with radius R_0 composed of a dispersal of chains of identical spheres. For the present, assume that all the chains have the same number of particles N and that they are all aligned parallel to the electric field E_0 as shown in Figure 6.22c. If $p_{eff}^{(1)}/N$

---

[3] Close consideration of the derivation provided here will convince the reader that the assumption of identical spheres is unnecessary.

is the moment per particle for a chain of N spheres normalized to p_0, then Equation (6.49) becomes

$$\frac{\varepsilon_{\text{eff}} - \varepsilon_1}{\varepsilon_{\text{eff}} + 2\varepsilon_1} = \varphi_1 K p_{\text{eff}}^{(1)}/N \tag{6.54}$$

or

$$\varepsilon_{\text{eff}}/\varepsilon_1 = \frac{1 + 2\varphi_1 K p_{\text{eff}}^{(1)}/N}{1 - \varphi_1 K p_{\text{eff}}^{(1)}/N} \tag{6.55}$$

Let us now restrict our attention to chains of identical conducting spheres, that is, $K = 1$, so that we can make use of the $p_{\text{eff}}^{(1)}/N$ values in Table 6.1 to calculate the effective permittivity of a heterogeneous medium comprised of aligned N-sphere chains. Figure 6.23 plots $\varepsilon_{\text{eff}}/\varepsilon_1$ as a function of the volume fraction φ_1 for chains of length $N = 1$ through 5. The strong influence of chain length on ε_{eff} is entirely analogous to the effect of eccentricity on the effective permittivity of heterogeneous mixtures of elongated particles (Rosensweig et al., 1981).

B.1 Distribution of chain lengths

It is natural to expect that there will be a distibution of chain lengths in a suspension of agglomerated particles. To represent this situation, assume that (i) the conductive spherical particles form long chains aligned parallel to the electric

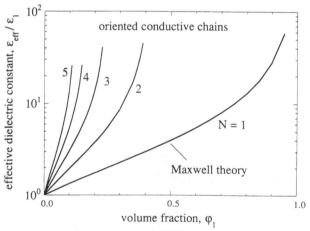

Fig. 6.23 Effective relative permittivity $\varepsilon_{\text{eff}}/\varepsilon_1$ for heterogeneous mixtures of oriented chains of perfectly conducting spheres: $N = 2, 3, 4,$ and 5. Result for a dispersal of isolated spheres ($N = 1$) is plotted for comparison.

field and (ii) the lengths are approximately characterized by a continuous distribution function $f(N)$, where

$$\int_1^\infty f(N)dN = 1 \tag{6.56}$$

The average length $\langle N \rangle$ is then

$$\langle N \rangle = \int_1^\infty Nf(N)dN \tag{6.57}$$

We now use this distribution of chain lengths and the Maxwell superposition principle with Equation (6.49) to express the effective permittivity of a mixture of aligned chains. The required expression for ε_{eff} is

$$\frac{\varepsilon_{\text{eff}} - \varepsilon_1}{\varepsilon_{\text{eff}} + 2\varepsilon_1} = \frac{\varphi_1}{\langle N \rangle} \int_1^\infty p_{\text{eff}}^{(1)} f(N)dN \tag{6.58}$$

Calculated values for $p_{\text{eff}}^{(1)}$, divided by N, are available for $N \le 7$ in Table 6.1. The log–log plot of these calculated results in Figure 6.12 suggests the simple power law: $p_{\text{eff}}^{(1)} = cN^3$ for large N. Then, Equation (6.58) reduces to

$$\varepsilon_{\text{eff}}/\varepsilon_1 = \frac{1 + 2\varphi_1 c \langle N^3 \rangle / \langle N \rangle}{1 - \varphi_1 c \langle N^3 \rangle / \langle N \rangle} \tag{6.59}$$

where $\langle N^3 \rangle$ is the average of N^3 for the chains in the suspension under the restriction that $\langle N \rangle \gg 1$.

B.2 *Distribution of chain orientations*

The assumption that all chains are aligned parallel to the electric field may be inappropriate for colloidal suspensions of small particles if Brownian motion competes successfully with the electrical alignment torque to randomize the chain orientations. As a simple example of how particle alignment influences the effective permittivity of the mixture, consider a dispersal of rigidly coupled two-sphere chains as depicted in Figure 6.22d. Assume a general axisymmetric orientational distribution $f(\theta)$ of these chains about the imposed electric field vector $\bar{E}_0$, where θ is the angle between the axis of the doublet and the field vector. The distribution $f(\theta)$ is normalized by an integration over a hemispheric surface.

$$\int_0^{\pi/2} f(\theta)2\pi\sin\theta d\theta = 1 \tag{6.60}$$

We now employ Equation (6.24) to obtain the dipole moment contribution parallel to the electric field. The expression for this moment, normalized to the moment of a single conductive sphere p_0, depends upon the angle θ.

$$\begin{aligned}
p_{\text{eff}}^{(1)}(\theta) &= p_{\text{eff}}^{(1)}/p_0 \\
&= \zeta(3)[8\cos^2\theta + 3\sin^2\theta]/2
\end{aligned} \tag{6.61}$$

Invocation of the superposition principle to sum the contributions of all chains requires an integration over the angle θ between 0 and $\pi/2$. The result is

$$\frac{\varepsilon_{\text{eff}} - \varepsilon_1}{\varepsilon_{\text{eff}} + 2\varepsilon_1} = \frac{\varphi_1}{2}\int_0^{\pi/2} p_{\text{eff}}^{(1)} f(\theta)2\pi\sin\theta d\theta \tag{6.62}$$

Equation (6.62) can be tested readily against the results of other analyses for the special case of randomly oriented doublets. In this case, there is no preferred alignment so that the distribution function is a constant with respect to solid angle. Using Equation (6.60), we obtain $f(\theta) = (2\pi)^{-1}$. Substituting this constant value for $f(\theta)$ and Equation (6.61) into (6.62), we integrate to obtain

$$\frac{\varepsilon_{\text{eff}} - \varepsilon_1}{\varepsilon_{\text{eff}} + 2\varepsilon_1} = [7\zeta(3)/6]\varphi_1 \cong 1.402\,\varphi_1 \tag{6.63}$$

This result agrees with the value for the dimensionless polarizability coefficient obtained by an analysis based on expansion of the Clausius–Mossotti function using integrals of excess cluster polarizabilities (Bedeaux et al., 1987).

6.6 Conclusion

In this chapter, we have explored the particle interactions that lead to chain formation in suspensions of conductive, dielectric, and magnetizable particles. We have found that these interactions become very strong for particles with spacing less that one-tenth of the radius ($\delta < R/10$) if the relative permittivity ($\varepsilon_2/\varepsilon_1$), conductivity ($\sigma_2/\sigma_1$), or permeability ($\mu_2/\mu_1$) of the particles is larger than approximately 10. Chain formation, an important phenomenon widely observed whenever ensembles of particles are subjected to an electric or magnetic field, is a direct consequence of the great strength of these interactions. The force interactions responsible for particle chain formation are the subject of the final chapter of this book.

7

Force interactions between particles

7.1 Introduction

In Chapter 6, we presented simple models for the field-induced interactions of linear chains consisting of dielectric, conductive, or magnetic spheres subjected to uniform electric or magnetic fields. In this chapter, we build upon this foundation, using, where appropriate, electrical images or effective linear multipoles to estimate interparticle force magnitudes between particles in electric and magnetic fields. Then, to validate the methodology and to establish its practical limits, we compare results to other theories and to available experimental measurements. Presentation of the force calculation models opens the door for consideration of chaining phenomena in a very broad context, including field-induced flocculation, the fibration of particles suspended in liquids, the strongly anisotropic electromechanics of granular media and powders, and the important topic of particle/surface adhesion.

7.2 Theory

The effective dipole method espoused in Section 2.4 of this book has as its basis the hypothesis that the forces and torques of electrical origin on a particle may be determined by replacing the actual particle by an equivalent point dipole that creates the same dipolar field as that attributed to the particle. Once the effective dipole has been identified, we may use it in Equations (2.36a,b) to determine the force and torque, respectively. Of course, these equations are only approximations, good only to the extent that the higher-order multipoles are negligible. Their neglect leads to inaccuracies of two types: more obviously, omission of the force interactions among the higher-order moments themselves, but also neglect of the strong influence of the higher-order terms on the dipole itself. For example, Figures 6.14 and 6.15 demonstrate how badly the dipole approximation fails at predicting even the net dipole moment p_{eff} in a chain of highly polarizable and closely spaced particles.

For reasons already cited in Section 6.1, we restrict attention here to chains of uncharged particles aligned parallel to the z-directed electric field. Given parallel alignment, all induced multipoles are linear and aligned with z. The logical starting point for calculating the net force between two particles is the interaction between two aligned linear multipoles of order m and n, located on the z axis at spacing Ξ and illustrated in Figure 7.1. The mutual force of attraction is proportional to the product of $p^{(m)}$ and $p^{(n)}$.

$$F_{m,n} = (-1)^{n+1} \frac{(m+n+1)! \, p^{(m)} p^{(n)}}{4\pi\varepsilon_1 \Xi^{m+n+2} m! n!} \tag{7.1}$$

The sign convention chosen for $F_{m,n}$ is that $p^{(m)}$ and $p^{(n)}$ attract one another when $F_{m,n} > 0$ and repel when $F_{m,n} < 0$. Equation (7.1) is an *action at a distance* generalization of Coulomb's Law for the force interaction between two linear multipoles aligned along a common axis.

Figure 7.2 shows spheres a and b of radii R_a and R_b aligned with their line of centers parallel to a uniform applied electric field E_0. All induced moments in both particles are linear multipoles aligned with z. The net mutual attractive force between the two particles is the double summation of all $F_{m,n}$ terms from Equation (7.1).

$$F_{\text{mutual}} = \frac{1}{4\pi\varepsilon_1} \sum_{m=1}^{\infty} \sum_{n=1}^{\infty} (-1)^{n+1} \frac{(m+n+1)! \, p_b^{(m)} p_a^{(n)}}{\Xi^{m+n+2} m! n!} \tag{7.2}$$

With perfect generality, all moments have been placed at the centers of the spheres. To use this formulation, we must solve for all the multipolar moments, $p_a^{(n)}$ and $p_b^{(n)}$, as functions of the imposed axial electric field E_0.

The relationships we need between these moments and the field, available from Section 6.3C, are repeated below for convenience.

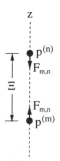

Fig. 7.1 Two interacting linear multipoles $p^{(m)}$ and $p^{(n)}$ aligned on the z axis at spacing Ξ with mutual force of attraction $F_{m,n}$.

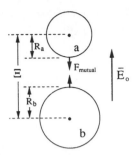

Fig. 7.2 Two spherical particles of radii R_a and R_b at center-to-center spacing Ξ with their line of centers parallel to the applied electrostatic field vector $\bar{E}_0$.

$$E_a = E_0 + \sum_{i=1}^{\infty} \frac{(i + 1)p_b^{(i)}}{4\pi\varepsilon_1 \Xi^{i + 2}} \tag{6.40a}$$

$$\frac{\partial^n E_a}{\partial z^n} = \sum_{i=1}^{\infty} \frac{(-1)^n (i + 1)(i + 2) \ldots (i + 1 + n)p_b^{(i)}}{4\pi\varepsilon_1 \Xi^{i + 2 + n}}, \; n \geq 1 \tag{6.40b}$$

with similar equations for E_b and its spatial derivatives in terms of $p_a^{(i)}$. Here, $E_a(z)$ and $E_b(z)$ are the axial electric fields at the centers of particles a and b, respectively. Equation (6.41) from Chapter 6 is replaced by a pair of relations.

$$p_a^{(n)} = \frac{4\pi\varepsilon_1 K^{(n)}R_a^{2n+1}}{(n-1)!} \frac{\partial^{n-1}E_a}{\partial z^{n-1}} \tag{7.3a}$$

$$p_b^{(n)} = \frac{4\pi\varepsilon_1 K^{(n)}R_b^{2n+1}}{(n-1)!} \frac{\partial^{n-1}E_b}{\partial z^{n-1}} \tag{7.3b}$$

There are practical limits on the accurate numerical computation of the multipoles. From Section 6.3B, we must anticipate serious convergence problems for small spacing (viz., $\Xi/R < 2.1$) and high relative permittivities, $\varepsilon_2/\varepsilon_1 > 10$ (or high permeabilities, $\mu_2/\mu_1 > 10$).

A. Force between conducting spheres

The special case of perfectly conducting interacting spheres is interesting because it represents the upper limit of particle polarizability. Somewhat ironically, we may not use multipolar expansions to treat perfectly conductive spheres at close spacing because of the convergence problem. Fortunately, we have recourse to the method of images already exploited in Section 6.2A.

Consider two identical perfectly conducting spheres of radius R and in contact. To proceed, we sum the coulombic interactions between all the image charges in both particles.

$$F_{\text{mutual}} = \frac{1}{4\pi\varepsilon_1}\sum_{m=1}^{\infty}\sum_{n=1}^{\infty}\left[q_m\left(\frac{q_n}{(d_m+d_n)^2}+\frac{q_n'}{(d_m+d_n')^2}\right)+\right.$$ (7.4)

$$\left.q_m'\left(\frac{q_n}{(d_m'+d_n)^2}+\frac{q_n'}{(d_m'+d_n')^2}\right)\right]$$

Figure 6.2b, with Equations (6.5a,b) and (6.6a,b), defines all charge and distance variables used in Equation (7.4). If care is exercised in taking the $R/D \rightarrow 0$ limit of this expression under the constraint that the field $E_0 = Q/2\pi\varepsilon_1 D^2$ remains finite, we obtain for the net attractive force

$$F_{\text{mutual}} = 4\pi\varepsilon_1 R^2 E_0^2 \sum_{m=1}^{\infty}\sum_{n=1}^{\infty}\frac{4mn-n^2-m^2}{(n+m)^4}$$ (7.5)

The published numerical evaluation of this double summation is incorrect (Jones, 1986b), probably because the summation fails the test of absolute convergence; however, a rigorously correct analytical evaluation has now been provided (Williams, 1993).

$$\lim_{M\to\infty}\sum_{m=1}^{M}\sum_{n=1}^{\infty}\frac{4mn-n^2-m^2}{(n+m)^4} = \zeta(2)-\zeta(3)-1/6$$ (7.6)

Equations (7.5) and (7.6) combine to yield an expression for F_{mutual} which may be validated by comparison to the solution of a related problem. Refer to Figure 7.3, which shows a perfectly conducting sphere in contact with a perfectly conducting plane and subject to a uniform normally directed electric field E_0. The *total* lifting force exerted on this particle consists of a vector sum of the image

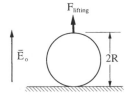

Fig. 7.3 A single perfectly conducting sphere resting on perfectly conducting plane with the applied normal electric field $\bar{E}_0$. $\bar{F}_{\text{lifting}}$ is the electrostatic lifting force defined by Equation (7.7).

force, identical to F_{mutual}, plus the coulombic force exerted by the imposed field E_0 upon the net charge induced in the sphere. This induced charge is the so-called Maxwell charge defined by Equation (6.7). Thus,

$$F_{\text{lifting}} = Q_a E_0 - F_{\text{mutual}}$$

$$= 4\pi\varepsilon_1 R^2 E_0^2 [\zeta(3) + 1/6]$$

(7.7)

Equation (7.7) is identical to an analytical solution obtained using bispherical coordinates to solve the boundary value problem and an integral of the Maxwell stress tensor to evaluate the total force (Lebedev and Skal'skaya, 1962).

When the two conducting particles are not in contact, we may return to the multipolar formalism of Equation (7.2).[1] Using Equation (7.3a,b) with $K^{(n)} = 1/n$ for conductive spheres,

$$p_a^{(n)} = \frac{4\pi\varepsilon_1 R^{2n+1}}{n!} \frac{\partial^{n-1} E_a}{\partial z^{n-1}} \quad \text{and} \quad p_b^{(n)} = \frac{4\pi\varepsilon_1 R^{2n+1}}{n!} \frac{\partial^{n-1} E_b}{\partial z^{n-1}}$$

(7.8)

Note that, because $R_a = R_b = R$, $p_b^{(n)} = (-1)^{n-1} p_a^{(n)}$.

Davis (1964) used bispherical coordinates to solve the general problem of two perfectly conducting, electrically charged spheres of different radii at arbitrary spacing in a uniform electric field. To calculate the interparticle force, he first obtained a series expression for the electric field between the spheres and then integrated the Maxwell stress tensor over a closed surface enclosing one of the particles. Unfortunately, a comparison of representative numerical results found in this paper to the multipolar formulation presented above is not a straightforward exercise.

B. Force between dielectric spheres

Despite its greater practical importance, the problem of dielectric particles in a uniform electric field (or the analogous problem of magnetic particles in a magnetic field) has received considerably less attention than that of perfectly conducting particles, due to the greater mathematical difficulty of the boundary value problem for dielectric particles. Analytical treatments using bispherical coordinates and integration of the Maxwell stress tensor are available (Davis, 1969; Love, 1975; Godin and Zil'bergleit, 1986), though easily used approximate expressions are difficult to extract from such analyses. The analysis of

---

[1] To avoid convergence problems, a restriction can be placed on the particle separation, namely, $\Xi/R > 2.1$ (or $\delta/R > 0.1$).

Chen et al. (1991), who used an expansion of spherical harmonics to calculate interparticle forces in infinite chains of dielectric spheres, is more closely related to the methods of this book.

Practical calculation of F_{mutual} requires truncating each series expression at some finite number of multipoles. We solve the resulting set of linear equations numerically for the moments within each sphere, $p_a^{(i)}$ and $p_b^{(i)}$, and then use these values in Equation (7.2) to calculate the force. Figure 7.4 plots the mutual force, normalized to $\varepsilon_1 E_0^2 \pi R^2$, versus the spacing δ/R at several values of the relative permittivity $\varepsilon_2/\varepsilon_1$. These curves level off and approach asymptotic limits as $\delta/R \to 0$ because, for any finite value of $\varepsilon_2/\varepsilon_1$, the magnitude of the higher-order multipoles becomes self-limiting as the particles approach contact. An equivalent physical interpretation of this result is that, as long as $\varepsilon_2/\varepsilon_1$ remains finite, the field in the gap between the particles approaches a finite limit as $\delta/R \to 0$.[2]

One may adapt the above analysis to the case of spherical magnetic particles in a magnetic field using the set of substitutions described in Appendix A, that is, $\varepsilon_1 \to \mu_1$, $\varepsilon_2 \to \mu_2$, $E_0 \to H_0$, etc. The magnetic interparticle force normalization is $\mu_1 H_0^2 \pi R^2$. The problem of perfectly conducting spheres, already discussed in Section 7.2A, serves as a model for the $\mu_2/\mu_1 \to \infty$ limit.

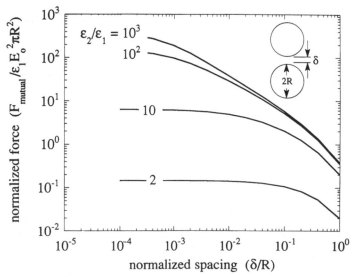

Fig. 7.4 Normalized attractive force F_{mutual} between two identical dielectric spheres of radius R and permittivity ε_2 in dielectric fluid of permittivity ε_1 plotted versus normalized spacing δ/R for several values of $\varepsilon_2/\varepsilon_1$.

---

[2] Compare this result to Equation (6.43), which is applicable only for chains of perfectly conducting particles, or the equivalent limits $\varepsilon_2/\varepsilon_1 \to \infty$ or $\mu_2/\mu_1 \to \infty$.

C. Current-controlled interparticle forces

When chains of slightly conductive dielectric particles are subjected to a DC electric field, current flows from particle to particle and the simple theory of linear multipoles loses its value in predicting interparticle forces. The DC current leads to a buildup of free electric surface charge on the adjacent surfaces of contacting particles, intensifying the electric field in the gap between the particles and creating very strong attractive electrostatic forces. Standard multipolar expansion methods are less convenient for calculating these forces because account must now be taken of nonlinear conduction mechanisms in the gaps between particles.

Nonlinear conduction models for electrostatic particle interactions were first proposed to explain the adhesion of fly-ash particulates to the collection plates in an electrostatic precipitator (McLean, 1977). For successful precipitator performance, the growing layer of collected particles must adhere to the wall so that particles are not re-entrained in the gas stream. The tenacious holding force experienced by precipitated particles is attributed to the current-controlled field intensification mentioned above. The accepted model for the electrostatic holding force does not employ multipoles but instead relies on analytic solutions for the Laplacian field using various approximations for current conduction from particle to particle (Moslehi, 1983).

According to one very simple model, the DC electric current flows only on the surface of the particles and passes from particle to particle within a small region where the field is intensified sufficiently to achieve gaseous breakdown. Assume that this interparticle current confines itself to a region of the gap where the field is constant and equal to E_{max}, as depicted in Figure 7.5. As the average (superficial) field E_0 is increased, the interparticle current and the field-limited volume within the gap grows. An approximate expression based on this model for the interparticle force is provided by Dietz and Melcher, 1978.

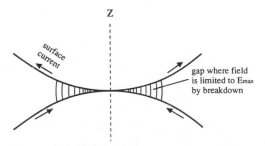

Fig. 7.5 Current-controlled field intensification in the gap between two closely spaced or touching dielectric particles with surface conduction. The field in the gap is limited to E_{max} by a gaseous breakdown mechanism (Dietz and Melcher, 1978).

$$F_{\text{mutual}} \cong 1.66\,\pi\varepsilon_1 R^2 E_{\text{max}}^{0.8} E_0^{1.2} \tag{7.9}$$

Though based on a very simple model, the form of Equation (7.9) is actually rather general because *any* nonlinear conduction process will tend to limit the gap field and weaken the dependence of the mutual force upon E_0 in a similar way. In fact, the simple power law, $F_{\text{mutual}} \propto E_0^\lambda$, with $1.1 \leq \lambda \leq 1.4$, provides a good prediction of the dependence of the force on the electric field when other conduction laws are used instead of the simple field-limited model of Dietz and Melcher (1978).

Current-controlled interparticle forces have yet to be measured directly on particle chains. Workers have instead measured the electric yield stress τ_E on stationary beds of dielectric particles stressed by an electric field (Johnson and Melcher, 1975; Dietz, 1976; Robinson and Jones, 1984). In particular, Robinson analyzed data from very diverse types of measurements and experimental conditions to extract a relationship between F_{mutual} and E_0, under the reasonable supposition that $\tau_E \propto F_{\text{mutual}}$ (Robinson, 1982). The yield stress data exhibit a relationship to the applied field E_0 that is slightly stronger than linear, consistent with $1.1 \leq \lambda \leq 1.4$, as predicted by theory.

7.3 Experiments

Harpavat's simple yet elegant experiments (1974) were first among a small handful of force measurements made upon chains of magnetic particles. He measured the relative magnitudes of the magnetic forces between 4.76-mm-diameter steel balls in linear chains of up to $N = 10$ particles resting against the surface of a permanent magnet, assembling chains and then pulling them apart to measure F_{mutual} and to locate the weakest joint in chains of different lengths. Some measurements on beads with nonmagnetic coatings were also reported. He then compared the results to predictions based upon a simple dipole approximation. Despite certain differences between experimental conditions and the assumptions of theory (having to do with the nonuniformity of the field) Harpavat nevertheless discovered the dipole theory to be a successful predictor of the weakest joint in particle chains.

A. Interparticle force measurements

Tan used an electronic force gauge to measure the mutual attractive forces between commercially available 3.34-mm-diameter chrome steel balls arranged in chains and planar arrays (Tan and Jones, 1993). The apparatus, shown in Figure 7.6, used a vertically mounted Helmholtz coil to produce a magnetic field

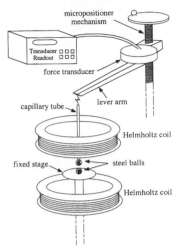

Fig. 7.6 Apparatus used to measure magnetic interparticle forces between individual elements in chains and layers of chrome steel balls (Tan and Jones, 1993).

H_0, uniform to less than $\pm 0.5\%$ over the length of the longest chains investigated. The electronic force gauge facilitated two types of measurements: (i) the attractive static force between particles at fixed spacings and (ii) the pull-away force required to break apart short chains at each of the different joints. The resulting data provide a test of the accuracy of multipolar force calculation methods.

Figure 7.7 shows normalized force data versus spacing for chains of $N = 2$ identical spheres. The points plotted along the left margin of this figure at $\delta/R = 10^{-5}$ are pull-away force measurements made on particle pairs in mechanical contact. The theoretical curve calculated from the multipolar analysis for $\mu_2/\mu_1 = 10^2$ (identical to the $\varepsilon_2/\varepsilon_1 = 10^2$ curve in Figure 7.4) illustrates that multipolar theory successfully predicts the attractive force. This success has important consequences because, as revealed next, the two-sphere interaction model serves as a good building block for the electromechanics of chains and layers.

The force data plotted in Figure 7.7 cover a range of almost a factor of 10 in the applied magnetic field (from $H_0 = 1.3$ to $10 \, \text{kA/m}$) and reveal no significant nonlinear effect. We examine the influence of magnetic nonlinearity on interparticle forces in Section 7.3C.

B. Measurements on longer chains and planar arrays

Figures 7.8a, b, and c display additional force measurements made on chains of $N = 3$ and $N = 4$ particles. Just as in Figure 7.7, the data plotted along the left margin at $\delta/R = 10^{-5}$ represent pull-away force measurements made on uncoated particles in contact. The theoretical curve in each of these plots is calculated from the multipolar analysis for the case of $N = 2$ spheres with $\mu_2/\mu_1 = 10^2$. As might be expected, the magnitude of the interparticle force increases modestly with N. Comparison of Figures 7.8b and c reveals another anticipated outcome that the mutual force is somewhat weaker at the joints closer to the ends of chains.

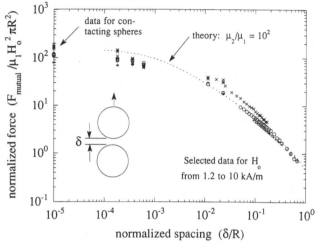

Fig. 7.7 Normalized attractive force between two 3.34-mm-diameter chrome steel balls versus normalized spacing δ/R (data from Tan and Jones, 1993). The data, obtained over the magnetic field intensity range from 1.2 kA/m to ~10 kA/m, show a modest nonlinear effect. The curve for $\mu_2/\mu_1 = 10^2$ is identical to the $\varepsilon_2/\varepsilon_1 = 10^2$ curve in Figure 7.4.

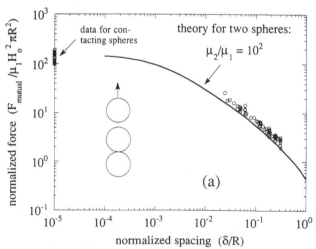

Fig. 7.8 Normalized attractive force between 3.34-mm-diameter chrome steel balls in linear arrays versus normalized spacing δ/R (data from Tan and Jones, 1993). (a) $N = 3$; (b) $N = 4$, force measured at joint between third and fourth particles; (c) $N = 4$, force measured at middle joint between second and third particles. [(b) and (c) are on facing page.]

To complement the measurements on chains, Tan also measured the attractive forces between single spheres and planar arrays (Tan and Jones, 1993). Data for a triangular array of three and a square array of nine identical spheres are provided in Figures 7.9a and b, respectively. For comparison's sake, we again superimpose the calculated curve for a linear chain of $N = 2$ spheres. The slight reduction of the holding force from the plotted curve is certainly due to demagnetization in the layer, the same effect evident in the $p_\perp$ data found in Figure 6.16. Experiments with other planar arrays are almost indistinguishable from these data.

Examination of Figures 7.8a, b, and c and 7.9a and b teaches us that the simple $N = 2$ theory correlates to within a factor of two for all particle arrays – chains

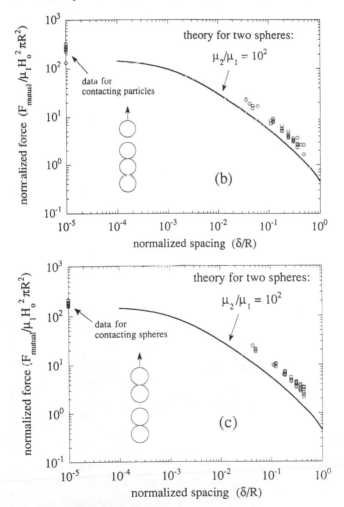

Fig. 7.8 Caption on facing page.

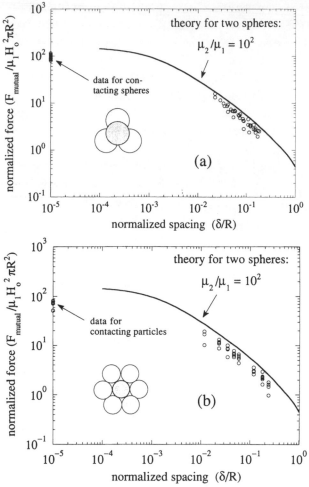

Fig. 7.9 Normalized attractive force between 3.34-mm-diameter chrome steel ball and planar arrays of the same balls versus normalized spacing δ/R (data from Tan and Jones, 1993). The theoretical curve is identical to the $\varepsilon_2/\varepsilon_1 = 10^2$ curve from Figure 7.4. (a) Triangular array of three balls; (b) hexagonal array of seven balls.

and layers – investigated. This outcome supports the principle that, with minor modification, the two-particle interaction serves as a basic building block for the electromechanics of more complex particle arrays.

C. Nonlinear effects

We have already examined the influence of material nonlinearity on the effective dipole moments of steel particle chains in Section 6.3D. In particular, Figure 6.20a shows that the effective dipole moment per particle, p_{eff}/N, exhibits a magnetic saturation effect clearly accentuated by chain length. At first glance, one would expect saturation to influence the interparticle forces even more strongly. This is not really the case, however, because of the way that the finite permeability limits the field in the gap when particles approach contact. Instead of saturation, it is hysteresis in the form of remanent magnetization (c.f., Figure 3.9b) that plays a larger role in interparticle forces. For example, reproducible force data at low magnetic field strengths for closely spaced particles are attainable only if care is taken to demagnetize the particles after each measurement (Tan and Jones, 1993).

The force measurements presented in Figure 7.7 were obtained at three different values of the magnetic field covering almost an order of magnitude of the applied field H_0. These data, corrected for the effects of remanent magnetism, reveal a rather small field-dependent effect that would not occur if the particles were magnetically linear. To document a larger saturation effect would require measurements at far stronger magnetic field strengths.

Demagnetization of particles after each measurement is necessary because, after a particle chain had been subjected to a DC magnetic field, the particles become permanently magnetized, exhibiting mutual attractive forces even with no applied field. Tan (1993) investigated this effect by subjecting pairs of initially demagnetized particles to a uniform DC magnetic field and then, with the field removed, measuring the force required to pull them apart. He then carefully reassembled the chains in the same orientation and measured the pull-away force several more times, without reapplication of the magnetic field. As shown in Figure 7.10, the force data obtained in this way generally correlate to the initial DC magnetizing field. More important, however, the pull-away force decreases dramatically after the chain is reassembled. In subsequent pull-away measurements, the pull-away force on any magnetized chain quickly settles to a final value. On the other hand, when this same experiment is repeated for particles not in intimate contact ($\delta/R \approx 10^{-4}$) during initial field application, the relative difference between the first and subsequent pull-away measurements almost completely disappears.

Such behavior is consistent with second quadrant operation of a permanent magnet actuator when the coercive force H_c is low, as is typical of chrome steels. The magnetization trajectory shown in Figure 7.11 represents the local $M(H)$ inside particles close to the contact regions as (1) the chain is first magnetized, (2) the field turned off, (3) the particles separated, (4) then rejoined, et seq. (Bates, 1961). The dashed lines represent the effective reluctance load lines of the rest of the magnetic circuit between the two spheres when the particles are in contact and when they are separated. This simple model explains why the interparticle holding force decreases so dramatically when the particles are first pulled apart.

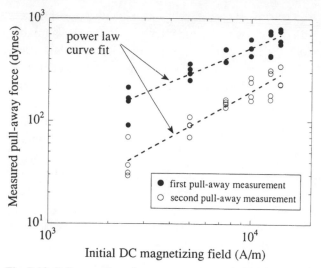

Fig. 7.10 Pull-away force between two magnetized 3.34-mm-diameter chrome steel balls with no applied field versus initial magnetizing field intensity. The second pull-away measurements are made with no reapplication of the field (Tan, 1993).

7.4 Electrostatic contributions to adhesion

The physics involved in the adhesion of a particle to a surface is very complex, involving short-range van der Waals forces, surface tension (if moisture is present), mechanical deformation of both the particle and the surface, and other phenomena, all additional to electrostatic forces. In this section, attention is limited to the electrostatic contributions to adhesion, a more general discussion being quite beyond the scope of this book. The prime purpose is to relate to adhesion the particle electromechanics already covered in Chapter 6 and in the preceding sections of this chapter.

A. *Phenomenological force expression*

Consider the basic problem of a charged spherical dielectric particle of radius R immersed in a fluid and resting on a planar dielectric surface, as shown in Figure 7.12a. For generality, we assume that the fluid, particle, and substrate have different permittivities, ε_1, ε_2, and ε_s, respectively. The particle, assumed to be perfectly insulating, has net charge q distributed uniformly over its entire surface. In addition, the particle is subjected to a uniform electric field E_0 oriented perpendicular to the planar surface. Following the lead of earlier workers (Goel and Spencer, 1975; Hartmann et al., 1976), we formulate the net electrostatic adhesion force in phenomenological form as the sum of three components.

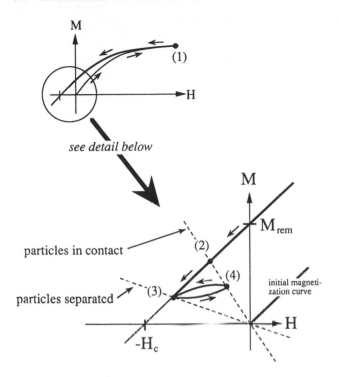

Fig. 7.11 Second-quadrant behavior of two permanently magnetized balls with low coercive force H_c. The important steps in the measurement procedure are: (1) magnetization of the chain, (2) turning off the field, (3) separation of the particles, and (4) reassembly of the chain. The dashed lines represent the load lines of the rest of the magnetic circuit between the two particles when in contact and separated.

$$F_{\text{adhesion}} = \alpha\left[\frac{q^2}{16\pi\varepsilon_1 R^2}\right] - \beta q E_0 + 4\gamma\pi\varepsilon_1 R^2 E_0^2 \tag{7.10}$$

The three terms in Equation (7.10), proportional to q^2, $q E_0$, and E_0^2, identify, respectively, with image, coulombic, and multipolar DEP force contributions. The term $q E_0$ carries a minus sign, indicating that, in xerographic applications, it is ordinarily a detachment force. As shown below, the coefficients α, β, and γ depend on the permittivities ε_1, ε_2, and ε_s.[3]

---

[3] Choice of α, β, and γ for the coefficients of the phenomenological components of the electrostatic adhesion force, while consistent with the literature, may confuse the unwary reader because these symbols have other meanings in prior chapters.

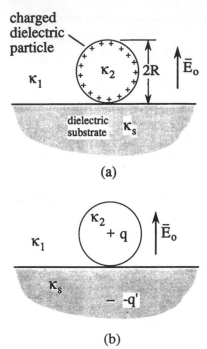

charged
dielectric
particle

κ_1

κ_2 2R $\bar{E}_o$

dielectric
substrate κ_s

(a)

κ_1

κ_2 $+ q$ $\bar{E}_o$

κ_s

$-$ -q'

(b)

Fig. 7.12 (a) Electrically charged particle of radius R and dielectric constant $\kappa_2 = \varepsilon_2/\varepsilon_0$ resting on substrate of dielectric constant κ_s. (b) Charged particle replaced by point charge q at center of particle showing image charge q'.

B. Image force contributions

The image charge force contribution (q^2) is positive because, except in the rather unlikely situation of $\varepsilon_s < \varepsilon_1$, the induced image charges have polarity opposite to q. The assumption that this charge is uniformly distributed on the surface of the particle is a critical feature of the model. If the particle's dielectric constant is the same as the fluid, that is, $\kappa_1 = \kappa_2$, then the spherically symmetric distribution of charge can be replaced by a point charge q at the particle's center, as shown in Figure 7.12b, and the (attractive) image force becomes

$$F_{image} = \frac{qq'}{16\pi\varepsilon_1 R^2}, \text{ where } q' = \frac{\kappa_s - \kappa_1}{\kappa_s + \kappa_1}q \qquad (7.11)$$

In Equation (7.11), q' is the magnitude of the negative image charge induced in the planar dielectric substrate (Weber, 1965). For convenience here, we introduce the dielectric constants of the substrate and the fluid, $\kappa_s = \varepsilon_s/\varepsilon_0$ and $\kappa_1 = \varepsilon_1/\varepsilon_0$, respectively. Note that if $\kappa_s/\kappa_1 \gg 1$, then $q' \to q$. The more general case of $\kappa_1 \neq \kappa_2$ must be accommodated by a correction factor α', which depends on all

Table 7.1 *Numerically calculated values of the image force correction coeffi-cient α' for various values of the dielectric constants κ_s and κ_2, with $\kappa_1 = 1.0$*

	$\kappa_s = 2.0$	3.0	4.0	5.0	6.0	∞
$\kappa_2 = 2.0$	1.076	1.120	1.152	1.172	1.188	1.33
3.0	1.128	1.212	1.268	1.312	1.348	—
4.0	1.164	1.280	1.368	1.432	1.484	—
5.0	1.192	1.336	1.448	1.536	1.604	2.20
10.0	—	—	—	—	—	4.80

Source: Goel and Spencer, 1975; Davis, 1969.

three dielectric constants and must be calculated numerically. Equation (7.11) takes the form

$$F_{\text{image}} = \alpha'(\kappa_s, \kappa_1, \kappa_2)\left(\frac{\kappa_s - \kappa_1}{\kappa_s + \kappa_1}\right)\frac{q^2}{16\pi\varepsilon_1 R^2} \tag{7.12}$$

In general, the necessary correction for the presence of the dielectric materials is modest, that is, $\alpha' \approx 1$ unless both $\kappa_2 > 2$ and $\kappa_s > 2$.

Electrostatic contributions to adhesion, including the image force, play an important role in the physics of xerographic copying, particularly in image development and photoreceptor cleaning processes. Table 7.1 gives numeri-cally computed values of α' for the case $\kappa_1 = 1$ over the ranges of κ_2 and κ_s typ-ical for xerographic toner and photoreceptor materials.

B.1 Particle interactions

The image force model used to derive Equation (7.12) neglects the influence of other nearby charged particles sitting on the dielectric surface. As illustrated in Figure 7.13, these neighbors induce their own image charges within the surface and, as a consequence, increase the net attractive image force experienced by each particle. We may estimate the strength of this enhancement by modeling the layer of toner particles as a regular array of dielectric particles, each possess-ing the same electric charge q. For any regular array, the lateral components of the force vector cancel and the net image force on each particle will be normal to the substrate and directed downward. The new result for the image force per particle is (Goel and Spencer, 1975).

$$(F_{\text{image}})_{\text{total}} = \frac{\alpha' q q'}{16\pi\varepsilon_1 R^2}\left[1 + \sum_i\sum_j \upsilon(r_{i,j})\left(\frac{1 + r_{i,j}^2}{\Xi^2}\right)^{-3/2}\right] \tag{7.13}$$

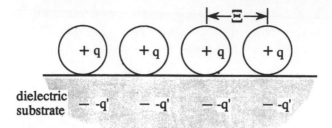

Fig. 7.13 Coulombic interactions among particle and image charges due to neighboring particles resting on a plane dielectric substrate.

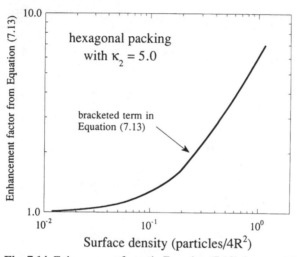

Fig. 7.14 Enhancement factor in Equation (7.13) due to neighbor interactions of identical particles in a hexagonal array plotted versus packing density. All data are calculated for a particle dielectric constant of $\kappa_2 = 5.0$ (Goel and Spencer, 1975).

Here, i and j are row and column indices signifying the locations of toner particles on the surface, while $r_{i,j}$ is the distance from the (i,j) image to the center of charge q, and $v(r_{i,j})$ depends sluggishly on $r_{i,j}$ and κ_2. The bracketed term in Equation (7.13) is the image force enhancement factor accounting for the neighbors' interactions. A calculation of this enhancement factor under the assumption of hexagonal packing is plotted in Figure 7.14 for $\kappa_2 = 5.0$. The force enhancement due to neighboring particle interactions is dramatic, approaching a factor of 10 for a hexagonal closely packed monolayer. A virtually identical curve results when other regular packing arrangements are assumed, indicating that nearest neighbor interactions are quite insensitive to the specific packing arrangement.

B.2 Nonuniform surface charge distribution

The electric image model of Figure 7.12 assumes that the charge is uniformly distributed on the surface of the particle. This spherical symmetry means that the distribution can be replaced by a point charge located at the center of the particle; however, there is no guarantee that the charge accumulated on dielectric particles by most natural processes will be symmetric. In fact, the triboelectric charging of toner particles in a xerographic copy machine can be expected to produce a charge distribution more like that depicted in Figure 7.15a. The qualitative and quantitative evidence for such a "patch" model is compelling (Hays, 1978; Lee and Ayala, 1985; Lee, 1986). It is of practical significance in electrophotography that this surface charge nonuniformity strongly influences the electrostatic adhesion.

Consider the very simple patch charge model shown in Figure 7.15b. The total particle charge q is divided into two portions: $q_p = fq$, located in a patch on the surface directly opposite to the planar substrate and $q_c = (1 - f)q$, located at the particle's center. We assume here that q_p and q_c have the same sign so that $0 \leq f \leq 1$. It is somewhat arbitrary to place q_c, which accounts for all surface

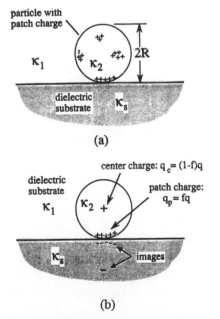

(a)

(b)

Fig. 7.15 The patch model for a charged toner particle. (a) Surface charge on a particle distributed in isolated patches. (b) Model for distribution of charge between patch of area A with charge $q_p = fq = \sigma_f A$ directly adjacent to substrate and remainder of charge $q_c = (1 - f)q$ located at the particle's center.

charge not included within the patch, at the center of the particle. We also assume that the charge in the patch is distributed uniformly over an area A.

$$q_p = A\sigma_f, \text{ where } A << 4\pi R^2 \tag{7.14}$$

where σ_f is uniform surface charge. As depicted in Figure 7.15b, both q_p and q_c induce images in the planar substrate.

To calculate the net force, we must take into account four different types of mutual interactions of q_p and q_c with the image charges they induce in the substrate.

(i) q_p' acting on q_c: $F_{(i)} = \dfrac{q_c' q_p}{4\pi\varepsilon_1 R^2} = -\dfrac{\xi q_c q_p}{4\pi\varepsilon_1 R^2}$

(ii) q_c' acting on q_c: $F_{(ii)} = \dfrac{\alpha' q_c' q_c}{16\pi\varepsilon_1 R^2} = -\dfrac{\alpha' \xi q_c^2}{16\pi\varepsilon_1 R^2}$

(iii) q_p' acting on q_p: $F_{(iii)} = \xi A \left[\dfrac{\sigma_f^2}{2\varepsilon_1}\right]$

(iv) q_c' acting on q_p: $F_{(iv)} = \dfrac{q_p' q_c}{4\pi\varepsilon_1 R^2} = -\dfrac{\xi q_c q_p}{4\pi\varepsilon_1 R^2}$

In the above, $\xi = (\kappa_s - \kappa_1)/(\kappa_s + \kappa_1)$ is the coefficient determining the magnitude of image charges in a dielectric plane[4] and α' is the same factor defined by Goel and Spencer (1975) and tabulated in Table 7.1. To calculate the net electrostatic adhesion force on the isolated particle, we now sum the above mutual attraction terms $F_{(i)}$, $F_{(ii)}$, $F_{(iii)}$, and $F_{(iv)}$.

$$F_{single} = \frac{\xi q^2}{2\pi\varepsilon_1 R^2}\left[(1-f)f + \frac{\alpha'(1-f)^2}{8} + \frac{3\sigma_f f}{4R\rho\left(\dfrac{q}{m}\right)}\right] \tag{7.15}$$

Equation (7.15) expresses the electrostatic particle adhesion force using parameters well-suited to the characterization of charged toner particles in a copier machine, including the total particle charge q, the surface charge density in the patch σ_f, the charge-to-mass ratio q/m, and the particle mass density ρ. For typical toners, $R \sim 5$ μm, $\sigma_f \sim 50$ nC/cm^2, $(q/m) \sim 20$ μC/g, and $\rho \sim 1.25$ g/cm^3. Note the strong dependence of each term in Equation (7.15) on the patch charge fraction f.

---

[4] This same factor ξ appears in Equation (7.11)

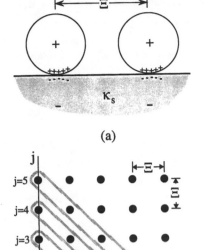

(a)

(b)

Fig. 7.16 (a) Interaction of toner particles having nonuniform charge distributions. (b) Top view of assumed cubic packing arrangement of particles on substrate.

Equation (7.15) is for a single isolated particle. If the density of particles adhering to the substrate is sufficiently high, then mutual interactions between particles must be taken into account. Refer to the diagram in Figure 7.16a, showing two adjacent particles with center-to-center spacing a. Because our interest is limited to the force acting normal to the substrate, some of the interaction terms may be ignored. For example, the normal component of the coulombic force interaction between the patch charge q_p on the leftmost particle and the center charge q_c on the rightmost particle in Figure 7.16a exactly cancels the similar interaction from right to left. The three contributions to the normal force acting on the rightmost particle are as follows:

$$F_{ij}^{A} = \frac{q_p' q_c R}{4 \pi \varepsilon_1 [R^2 + \rho_{ij}^2]^{3/2}}$$ (7.16a)

$$F_{ij}^{B} = \frac{q_c' q_c R}{2 \pi \varepsilon_1 [4R^2 + \rho_{ij}^2]^{3/2}}$$ (7.16b)

$$F_{ij}^C = \frac{q_c' q_p R}{4\pi\varepsilon_1 [R^2 + \rho_{ij}^2]^{3/2}} \tag{7.16c}$$

where the planar spacing parameter used in Equations (7.16a,b,c) is defined in Figure 7.16b.

$$\rho_{ij} = \Xi\sqrt{i^2 + (j-1)^2} \tag{7.17}$$

To calculate the net interaction force on the test particle, we must sum the contributions of all neighboring particles. A computationally effective way to perform the required double summations is to add the contributions of particles along diagonals, as depicted in Figure 7.16b. Using Equations (7.16a,b,c) and (7.17), we obtain

$$F_{\text{interaction}} = \frac{2\xi(1-f)q^2}{\pi\varepsilon_1 R^2}[fS_1 + (1-f)S_2] \tag{7.18}$$

with the following definitions

$$S_1 = \sum_{n=1}^{\infty}\sum_{m=1}^{n} \frac{1}{\left\{1 + \left[m^2 + (n-m)^2\right]\left(\frac{\Xi}{R}\right)^2\right\}^{3/2}} \tag{7.19a}$$

$$S_2 = \sum_{n=1}^{\infty}\sum_{m=1}^{n} \frac{1}{\left\{4 + \left[m^2 + (n-m)^2\right]\left(\frac{\Xi}{R}\right)^2\right\}^{3/2}} \tag{7.19b}$$

In the limit of cubic close packing ($\Xi = 2R$), these double summations have the following values: $S_1 \approx 0.24$ and $S_2 \approx 0.17$. The total electrostatic adhesion force is the sum of Equations (7.15) and (7.18).

$$F_{\text{total}} = F_{\text{single}} + F_{\text{interaction}}$$

$$= \frac{\xi q^2}{2\pi\varepsilon_1 R^2}\left[(1-f)f + \frac{\alpha'(1-f)^2}{8} + \frac{3\sigma_f f}{4R\rho\left(\frac{q}{m}\right)} + 4(1-f)[fS_1 + (1-f)S_2]\right] \tag{7.20}$$

The plots of F_{single}, $F_{\text{interaction}}$, and F_{total} versus f in Figure 7.17 teach two lessons: (i) charge localization on the surface of a particle dramatically affects the electrostatic adhesion force, and (ii) mutual interactions are strongest when charge localization is minimal, that is, when $f \approx 0$.

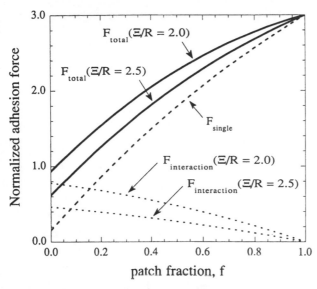

Fig. 7.17 Plots of F_{single}, $F_{interaction}$, and F_{total} in arbitrary units versus patch fraction f for $\Xi = 2R$ and other parameters: $\alpha = 1.22$ and $\sigma_f = 50$ nC/cm^2.

C. Detachment force contribution

The second term in Equation (7.10) has a negative sign, signifying that this elec-trostatic force contribution acts to detach the particle from the surface if $qE_0 > 0$. To obtain an expression for the enhancement factor β, we focus on the interac-tions of certain image charges and induced moments. This seemingly arbitrary partition is imposed because we are collecting only the terms proportional to qE_0. Any terms neglected here either depend on q' and have been incorporated into the first term of Equation (7.10), or depend on E_0^2 and will be incorporated into the third term of the same phenomenological equation.

Consider a spherical particle of radius R, dielectric constant κ_2, and electric charge q resting on a substrate of dielectric constant κ_s. The particle is under the influence of a uniform electric field E_0 normal to the surface. To estimate β, we approximate the detachment force by three terms.

$$\bar{F}_{detach} = \beta q \bar{E}_0 \approx q\bar{E}_0 + q\bar{E}_{image\ dipole} + (\bar{p} \bullet \nabla)\bar{E}_{image\ monopole} \qquad (7.21)$$

where $\bar{p} = 4\pi\varepsilon_1 R^3 K(\varepsilon_2, \varepsilon_1)\bar{E}_0$ is the dipole moment induced in the particle by the imposed field. The field quantities defined in Equation (7.21) are

$$E_{image\ monopole} = -q'/16\pi\varepsilon_1 R^2 \quad \text{and} \quad E_{image\ dipole} = p'/16\pi\varepsilon_1 R^3 \qquad (7.22)$$

with $q' = \xi q$, $p' = \xi p$, and the image coefficient $\xi = (\kappa_s - \kappa_1)/(\kappa_s + \kappa_1)$. Plugging

Equation (7.22) into (7.21) and comparing the result to the second term in Equation (7.10), we obtain

$$\beta \approx 1 + \frac{1}{2}\left[\frac{\kappa_s - \kappa_1}{\kappa_s + \kappa_1}\right]\left[\frac{\kappa_2 - \kappa_1}{\kappa_2 + 2\kappa_1}\right]$$ (7.23)

For the dielectric constant values typical of toners and photoreceptor materials, β never deviates significantly from unity, indicating that the enhancement of the detachment force by the substrate layer is not very important in copiers.

D. Induced moment contributions

The third term in Equation (7.10), proportional to E_0^2, arises from the multipolar moments induced in the particle by the imposed field. These moments create their own images in the dielectric substrate, and it is the interactions among them and their images that lead to all the E_0^2 terms. The simplest possible model for the induced field contribution to the electrostatic adhesion force involves the interaction of the dipole induced by E_0 in the particle with its own image in the dielectric substrate. Starting from the DEP force equation and employing the expression for the electric field due to the image dipole, we obtain

$$\bar{F}_{\text{ind. moment}} \approx (\bar{p} \bullet \nabla)\bar{E}_{\text{image dipole}} = -\frac{3pp'}{32\pi\varepsilon_0 R^4}\hat{z}$$ (7.24)

which is identical to Equation (7.1) with $n = m = 1$. Now, we can use Equation (7.24) to obtain an expression for the coefficient γ of Equation (7.10).

$$\gamma = \frac{3}{8}\left[\frac{\kappa_s - \kappa_1}{\kappa_s + \kappa_1}\right]\left[\frac{\kappa_2 - \kappa_1}{\kappa_2 + 2\kappa_1}\right]^2$$ (7.25)

In Section 6.3B, we learned that neglecting the higher-order multipoles leads to significant inaccuracy in estimating the effective dipole moment under certain conditions. Consequently, we should expect that Equation (7.25) seriously underestimates $\gamma(\kappa_1, \kappa_2, \kappa_s)$ when $\kappa_2/\kappa_1 > 4.0$.

E. Generalized model of Fowlkes and Robinson

We summarize below the analysis of Fowlkes and Robinson (1988), who adopted the method of effective multipoles to calculate the electrostatic adhesion force for the special case of a charged dielectric sphere resting on a *conductive* plane surface. They assumed that the charge, uniformly distributed on the particle's surface, can be represented by a point charge q at the center of the sphere. See Figure 7.18a. With $n = 0$ reserved for the monopole, that is, $p^{(0)} = q$, Equation (6.41) gives the needed relationships between the induced moments $p^{(n)}$, all located at the center of the sphere, and the electric field $E_z(z)$.

$$p^{(n)} = \begin{cases} q, & n = 0 \\ \dfrac{4\pi\varepsilon_1 K^{(n)} R^{2n+1}}{(n-1)!} \dfrac{\partial^{n-1} E_z}{\partial z^{n-1}}, & n > 0 \end{cases} \tag{7.26}$$

We can relate $E_z(z)$ to the imposed field and the image moments located in the conductive plane $p'^{(n)}$ using Equation (6.40b) with $\Xi = 2R$. Note that here we must include the contribution due to the monopolar image $p'^{(0)} = -q$.

$$E_z(z) = E_0 + \sum_{n=0}^{\infty} \frac{(n+1) p'^{(n)}}{4\pi\varepsilon_1 z^{n+2}} \tag{7.27}$$

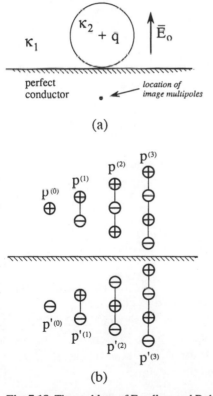

(a)

(b)

Fig. 7.18 The problem of Fowlkes and Robinson (1988). (a) Charged dielectric sphere of radius R resting on a conductive plane and in uniform normal electric field $\bar{E}_0$. (b) Relationship of the induced linear multipoles of the particle $p^{(n)}$ to their corresponding images $p'^{(n)}$.

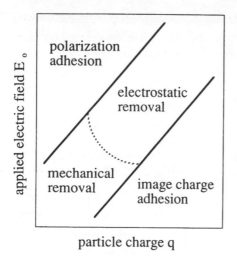

applied electric field E_o

polarization
adhesion

electrostatic
removal

mechanical
removal

image charge
adhesion

particle charge q

Fig. 7.19 Electrostatic adhesion regimes as functions of particle charge q and detachment field E_0. Calculated curves for sets of parameters relevant to toner particle detachment in xerographic copiers may be found elsewhere (Fowlkes and Robinson, 1988).

Because the planar surface on which the particle rests is conductive, all the image multipoles $p'^{(n)}$ can be related easily to the induced multipoles of the particle $p^{(n)}$, as illustrated in Figure 7.18b.

$$p'^{(n)} = (-1)^{n+1}p^{(n)}, n \geq 0 \tag{7.28}$$

One can solve Equations (7.26), (7.27), and (7.28) numerically for the moments $p^{(n)}$, though truncation of the series in Equation (7.27) is necessitated.

The net adhesive force on the charged particle is

$$F_{\text{adhesion}} = -qE_0 + \frac{1}{4\pi\varepsilon_0}\sum_{n=0}^{\infty}\sum_{k=0}^{\infty}\frac{(-1)^{n+k}(n+k+1)!p^{(n)}p^{(k)}}{n!k!(2R)^{n+k+2}} \tag{7.29}$$

In this formulation, the separate identity of the three phenomenological terms in Equation (7.10) is obscured because all moments depend on both q and E_0.

To illustrate the interplay between charge q and detachment field E_0, Fowlkes and Robinson devised the graphic presentation shown in Figure 7.19 for their calculated results. They assumed the existence of an additional nonelectrostatic force F_A and then defined four limiting physical regimes.

polarization adhesion   Here, the electric field is strong and the charge q is small so that the E_0^2 term in Equation (7.10) dominates and must be overcome to achieve mechanical particle removal.

image force adhesion In this case, the electric field is weak and the charge q is large so that the q^2 term in Equation (7.10) dominates. Mechanical particle removal requires overcoming the strong image force.

mechanical removal Now, both the charge and the applied electric field are small so that it is the nonelectrostatic force F_A that must be overcome for mechanical removal.

electrostatic removal Here, neither the q^2 nor E_0^2 terms dominate and the coulombic force qE_0 is strong enough so that the electric field can detach the particle from the surface.

Calculations performed using values of q and E_0 typical of the cleaning and development processes in xerographic copiers and printers reveal a result not previously appreciated that multipolar force contributions proportional to E_0^2 can be very important (Fowlkes and Robinson, 1988).

An underlying assumption of Equation (7.29) is that the effective center of the charge q is at the center of the particle. As already revealed in Section 7.4B, the evidence suggests that this assumption is not accurate for triboelectrically charged toner particles (Hays, 1978; Lee and Ayala, 1985; Lee, 1986). Hays and Wayman (1989) measured the detachment force and charge on single particles resting on the lower of two parallel electrodes stressed by a DC voltage. They observed detachment events as well as features of the nonuniformity of the surface charge by monitoring the charge induced on the lower electrode. In particular, they observed oscillatory components in the induced charge signal, which they successfully attributed to *permanent* dipole and quadrupole moments of the toner particles. Their experimental data and analysis strongly support the view that electrostatic forces, enhanced by charge nonuniformity, dominate other toner adhesion mechanisms in copiers.

7.5 Mechanics of chains and layers

Chapters 1 through 5 of this book focus on the forces and torques exerted by electric or magnetic fields upon individual particles and short chains. Chapters 6 and 7 build on these results, exploiting the method of effective multipoles to develop usable models for the attractive or repulsive forces between pairs of particles under rather idealized conditions. At least conceptually, pairwise particle interactions may serve as the building blocks for construction of electromechanical models for more complex ensembles such as chains and three-dimensional layers. On the other hand, most particulate technologies involve large ensembles of strongly interacting particles, necessitating more general interaction models. In the past few years, such technologies as electrorheological (ER) fluids, electrofluidized beds, magnetic brush copiers and printers, and magnetic powder couplings have provided a forceful incentive for concerted experimental and theoretical work on the electromechanics of powders and suspensions. In the concluding sections of this last chapter, we describe some models for systems of interacting particles and, along the way, identify some of the important problems requiring solution.

A very good example of a system where particle electromechanics are important is the magnetic brush xerographic copier. In typical copiers, magnetizable carrier particles (~100 μm in diameter) deliver charged toner particles (~10 μm in diameter) to a light-sensitive belt or drum to "develop" a latent electrostatic image. The carrier transport mechanism is called *magnetic tumbling* and it is depicted in Figure 7.20a. Toner-loaded carrier is cycled through the xerographic engine using a system of rollers, each of which has a set of internally mounted permanent magnets. On these rollers, the carrier particles form chains parallel to the magnetic field and, as the field alignment changes, the chains respond by tumbling end over end. In open sectors of a transport roll, the chain structure produces an almost fluffy, brush-like structure of low density. But where rollers come into close proximity to the photoreceptor or to metering structures (doctor blades), the brush is compacted and the powder mechanics is more complex (Jones et al., 1987). Clearly, a general model for the magnetic brush must go

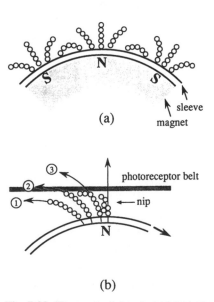

(a) sleeve magnet

(b)

Fig. 7.20 The magnetic brush. (a) Magnetic tumbling of a sparse brush. Either the sleeve can rotate with the underlying magnets stationary or the sleeve can remain stationary while the magnets rotate. (b) Chain breakage due to interference with photoreceptor belt.

beyond simple chain structures to deal with the far more difficult problem of dense, randomly packed beds of magnetized particles.

A. Chains of magnetizable particles

The problem of isolated or weakly interacting chains is the proper starting point for an investigation of magnetic brush mechanics. Several papers, each aimed specifically at modeling the magnetic brush, resort to the dipole approximation for chains of identical magnetic spheres aligned parallel to uniform (Harpavat, 1974) and nonuniform (Alward and Imaino, 1986) magnetic fields. The focus of these papers is identification of the weakest link in long chains of magnetic particles; in both, the authors invoke the dipole approximation to calculate the *relative* values of the attractive forces between adjacent particles in chains of various lengths. Unfortunately, as pointed out by Yarmchuk and Janak (1982) and clearly illustrated in Table 6.3, the accuracy of the dipole approximation and its predictions are open to question for particles with high relative permeability.

One analysis, founded on a generalization of Harpavat's dipole approximation, investigates the behavior of large arrays of chained magnetic particles (Paranjpe and Elrod, 1986). The distinguishing feature of this model is that it accounts in approximate fashion for the interactions among particles in adjacent chains and makes no a priori assumption about the alignment of individual dipole moments to either the applied magnetic field or the axis of the chain. Evaluation of the dipole moment vector for each particle proceeds by straightforward superposition of the effects of the nearest neighbors and the nonuniform applied magnetic field. Subject to the constraint of a fixed number of particles per unit area in the brush, the optimum chain length can be predicted by minimization of a magnetostatic potential energy associated with nearest neighbor interactions. Using this method, it is also possible to describe the fracture of chains as the magnetic brush enters the nip and starts to interfere with the photoreceptor belt, as shown in Figure 7.20b.

B. Mechanics of particle beds

In magnetic brush copier engines, dense random packing occurs when developer is forced through the development nip. Usually, the particles emerge from the nip as a compacted layer, only to break up abruptly and reform the brush structure at some fixed location that depends on the orientation and strength of the magnetic field. No model exists that can predict the location downstream from the nip for this spontaneous disruption. Chain models are not useful here because each particle in the densely packed layer is strongly coupled to its

neighbors. A second example of a packed magnetically coupled particle bed is
the *stagnant layer* observed in single-component magnetic brush copier engines
(Jones, 1988). The stagnant layer forms underneath the active tumbling brush
when a thick layer of low-permeability magnetic toner particles is deposited on
a magnet roll.

In an effort to model the electromechanics of packed magnetic powders,
Whittaker used a point dipole approximation to estimate the *internal magnetic
cohesion* for regular three-dimensional arrays of magnetic particles (see App. B
of Jones et al., 1987). This analysis consists of a summation of the dipole interaction energies for a semi-infinite line of identical magnetic dipoles interacting
with all the dipoles on the other side of the half-space. See Figure 7.21a. The
electromechanical cohesive stress results from invoking the virtual work concept to take the appropriate spatial derivative of this interaction energy.

An interesting prediction of this model is that the magnetic cohesion is
negative in planes parallel to the externally applied magnetic field. Such a

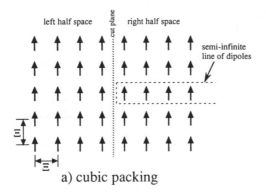

a) cubic packing

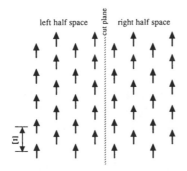

b) hexagonal close packing

Fig. 7.21 Regular array of identical dipoles used to calculate magnetic cohesion in beds
of magnetized particles. (a) Cubic array giving negative cohesion in planes parallel to
applied magnetic field (Jones et al., 1987). (b) Hexagonal close-packed array.

prediction is in fact not surprising because, for particles arranged in a cubic lattice, the repulsive interactions of adjacent side-by-side dipoles clearly dominate. It should also be clear, however, that the predicted cohesive stress will be strongly dependent upon the assumed packing. For example, the regular hexagonal packing shown in Figure 7.21b gives a much smaller (though still negative) cohesion.[5] To date, no direct experimental evidence for negative magnetic cohesion in beds of magnetizable particles has been reported. Nevertheless, Whittaker's theory predicts with reasonable accuracy the thickness of the stagnant layer observed in the tumbling of low-permeability magnetic toners (Jones et al., 1987). The theory also agrees with the observation that no stagnant layer occurs in high-permeability carriers.

The electromechanics of closely packed ensembles of magnetizable particles subjected to a magnetic field is relevant to other important applications such as magnetic powder couplings and magnetofluidized beds. Refer to Section 7.6D for further discussion of these examples.

7.6 Discussion of applications

The strong attractive particle interactions responsible for chaining exist in almost any physical situation where polarizable or magnetizable particles are subjected to electric or magnetic fields. Many engineered particulate systems beneficially exploit these phenomena. There are also situations where strong particle interactions have a deleterious effect. In this final section of the book, we review some well-known examples and proposed applications where particle electromechanics dominate. An underlying purpose of this summary is to make the strongest possible case for the commonality of the phenomena and to argue that, in many cases, a systematic treatment of particle interactions using the method of effective multipoles can provide a successful predictive model.

A. Electrofusion of biological cells

Electric field–mediated cell fusion has established itself as an important technique in biomedical research. The ability to create fused hybrid cells combining the attributes of different cell lines and types may facilitate genetic research, leading to the development of new drugs and therapies. One well-known protocol for electric field–induced fusion of biological cells depends upon an AC electric field (up to ~100 V/cm at ~500 kHz) to form chains of cells that collect on electrode structures suspended in an aqueous medium (Zimmermann and Vienken, 1982).

---

[5] When random packing and the strong influence of higher-order multipoles are taken into account for highly permeable particles, the reality of negative cohesion itself is thrown into doubt.

After chain formation, the AC field is removed briefly and one or more DC pulses are applied immediately before the cells can drift apart.[6] The pulses, of magnitude ~500 V/cm and duration ~10 µS, temporarily rupture the membranes of adjacent cells in their region of contact. The lipid bilayer then presumably reassembles itself into a single envelope holding the contents of both cells. Immediate reapplication of the AC electric field after the DC fusion pulses tends to improve the yield of such hybrids. It is likely that the interparticle force, responsible for initial chain formation (Stenger and Hui, 1988; Stenger et al., 1991), also beneficially influences the fusion process after application of the DC pulse.

B. Chaining in electrorheological fluids

The discovery of the electrorheological (ER) effect is generally attributed to Winslow (1949), who discovered that stabilized colloidal suspensions of certain particles in insulating dielectric liquids exhibit electric field–dependent viscosity. Technology stemming from this discovery promises new classes of high-speed, voltage-controlled dampers, shock absorbers, and torque transmissions. Under DC electrical stress, a typical ER fluid takes on the rheological attributes of gelatin; removal of the field returns the fluid immediately to normal Newtonian behavior. Under AC stress, the ER effect is strongly dependent on electric field frequency and rapidly diminishes for frequencies above ~10^2 Hz.

According to the generally accepted *fibration* theory, the ER effect is due to particle chains induced by the electric field, which form a fibrous web in the fluid volume and increase the effective viscosity (Halsey, 1992). With the recognition that the ER effect is explainable in terms of electromechanical interactions and that the details of the polarization mechanism are not really crucial (Block et al., 1990), many researchers have now turned their attention to models for chains. At least two different mechanistic theories have been proposed for how interparticle forces and chain alignment torques increase the effective viscosity (Halsey and Martin, 1993). According to one theory depicted in Figure 7.22a, initially intact chains break along some central shear plane while clinging to each electrode (Klingenberg and Zukoski, 1990). This model postulates that the electroviscous effect results from constriction of the shear zone by these broken chains. The other theory, illustrated in Figure 7.22b, proposes that free-floating chains in the shear flow experience an electrical alignment torque that impedes their rotation (Halsey et al., 1992). The chain alignment mechanism here is of course the same as that identified in Section 6.2C for linear particle chains.

---

[6] A simple alternate protocol is to form oriented chains in a suspension containing the two cell species with a uniform AC electric field and then to apply the DC fusion pulse (Teissie et al., 1982).

Direct calculations of interparticle forces have been performed using variants of the multipolar expansion method (Chen et al., 1991). They report strong dependence of the local electric field in the gap region between adjacent particles upon ε_1, ε_2 (the fluid and particle permittivities, respectively) and δ (particle spacing). In this calculation, they ignore particle and fluid conductivities. The predicted field intensification is sufficiently high to suggest nonlinear electrical conduction or breakdown in the gap between particle chains similar to packed particle beds in electrostatic precipitators (c.f., Section 7.2C). In ER fluids, one expects that nonlinear conduction might have an important effect on both heating and interparticle forces. Another recent work uses finite element analysis to calculate the forces between particles and the resulting effective shear coefficient (Davis, 1992). In this calculation, the ER fluid is modeled as a regular array of dielectric particles. Just as do multipolar calculations, this approach suffers convergence problems for particles at very small spacings; however, the model is probably quite appropriate in the slip zone depicted by Figure 7.22a.

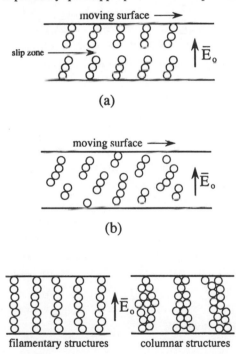

(a)

(b)

filamentary structures   columnar structures

(c)

Fig. 7.22 Mechanistic models for the electroviscous effect of electrorheological fluids. (a) Fractured chain remnants adhere to electrodes, creating a thin slip zone in the middle (Klinkenberg and Zukoski, 1990). (b) Shortened chains fill the inter-electrode space where they resist rotation due to the shear flow (Halsey et al., 1992). (c) Coarsening of chains into columns in an ER fluid.

The finding that the fibration structures in ER fluids are not simple linear chains, but rather thick columns, suggests that fibration must evolve in two distinct stages on two different time scales, namely (i) initial rapid formation of thin filaments and (ii) subsequent collection of these filaments into columns (Halsey, 1992). Refer to Figure 7.22c. The greater resiliency of columnar structures to stretching and bending in a fluid shear field has not escaped notice. A generalization of the stability analysis already performed for linear chains of magnetic particles (Paranjpe and Elrod, 1986) might help in determining whether the columnar structures elongate smoothly or break and reform during continuous shear.

Researchers seeking to improve ER fluid performance have investigated many different combinations of particles and liquids with wide-ranging conductivities. As recognized in Chapter 6 of this book, strong chaining in a DC field requires that the ratio of the effective particle-to-liquid conductivity be large compared to unity, that is, $\sigma_2/\sigma_1 \gg 1$. At the same time, it is necessary that both σ_2 and σ_1 be sufficiently low so that Joule heating does not become a problem. Thus, two common features of all materials packages exhibiting a measurable ER effect are (i) very low conductivity of the liquid vehicle σ_1 and (ii) moderately low conductivity of the particles σ_2. Early experiments with ER fluids revealed that the effect is enhanced if the particles contain sufficient moisture to guarantee the condition $\sigma_2/\sigma_1 \gg 1$ without causing heating problems. Unfortunately, ER fluids exhibiting such moisture dependence have neither good shelf life nor adequate long-term operating performance, no doubt because moisture content is uncontrollable. DEP levitation measurements on 50-mm-diameter glass beads in silicon oil show that moisture has a very strong effect on low-frequency polarizability (Tombs and Jones, 1993). Because the polarizability controls the interparticle forces responsible for chaining, the link between these observations and the documented sensitivity of ER fluids to moisture seems clear.

One solution to the moisture sensitivity problem in ER fluids is to develop particles with appropriate inherent bulk electrical conduction properties such as semiconductive polymers (Block et al., 1990). A rather novel idea is to employ conductive particles covered with a thin insulating layer, the advantage being that ER fluids formulated with such particles work in AC fields. Test fluids made with oxide-coated aluminum particles exhibit an ER effect that is actually stronger in AC fields than in DC (Inoue et al., 1992).

C. *Electrofluidized and electropacked beds*

Workers have been investigating the influence of electric fields on the electromechanics of beds of semi-insulating dielectric particles since McLean's discovery (1977) that interparticle forces strongly influence the cohesion of precipitated fly-ash particles in electrostatic precipitators. These electrostatic forces inhibit re-entrainment of particles into the gas stream and thus have a beneficial influence on collection efficiency. Current-controlled interparticle forces create very strong particle cohesion (Dietz and Melcher, 1978) not readily described using the simple multipolar interaction model. In this section, we identify other applications where electric fields are used to influence the mechanics of moving or fluidized beds of insulating dielectric particles.

An electric field alters dramatically the conditions of fluidization in a bed of dielectric particles. In particular, the flow rate required to achieve fluidization increases, the appearance of "bubbles" changes, and particle mixing is suppressed. If the electric field is strong enough, the particles become locked into a network of chain-like structures that prevents fluidization altogether. This "frozen" state, achieved where the field-induced interparticle forces become so strong that the mixing and turbulence of fluidization cannot overcome them, has possible applications in a variety of particulate flow control technologies (Johnson and Melcher, 1975; Martin et al., 1990). Another application for the electromechanical forces in moving beds of dielectric particles is the *electrospouted bed*, which uses a set of electrodes mounted near the gas inlet to control particle recirculation (Talbert et al., 1984).

The electromechanical behavior of electrofluidized and electrospouted beds contrasts with *electric suspensions*, which ordinarily do not involve significant chain formation. Rather, the particles, which are usually conductive, become electrically charged as soon as they come into contact with the electrodes. The resulting coulombic lifting force on the particles achieves the suspension effect. Electric suspensions of conductive particles in liquids (Dietz and Melcher, 1975) and in air (Colver, 1983) have been investigated for potential applications in heat and mass transfer augmentation.

Another proposed application for particle electromechanics involves using an AC electric field to improve the strength of clay molds before firing of earthenware and ceramics (Lo et al., 1992). Presumably, this effect is attributed to the enhancement of particle-to-particle bonding by the strong interparticle forces. Modeling the forces between individual kaolin particles in wetted clays is complicated by two factors. First, the particles are very tightly packed; second, an electrical double layer exists at the particle–liquid interface. This double layer, with its diffuse ionic charges extending out into the surrounding fluid, makes identification of effective multipoles and calculation of interparticle forces a difficult proposition.

D. Magnetopacked and magnetofluidized beds

The strong attractive forces exhibited by magnetizable particles in uniform (or nonuniform) magnetic fields have many important applications in addition to the magnetic brush xerographic engine discussed in Section 7.5. Analogous to the behavior of a bed of dielectric particles in an electric field, magnetizable particles also can be made to "freeze" into a rigid mass by application of a magnetic field. This effect has been exploited in control valves for magnetizable solids (Yang et al., 1982) and the magnetospouted bed (Jones et al., 1982). For many years, *magnetic clutches* using ferromagnetic granules have been in commercial use, including dry (Grebe, 1952) and liquid-based couplings (Jones, 1953). A reluctance model for predicting the yield conditions in such torque transmissions has been developed; it accounts for the particulate nature of the coupling through the very strong influence of the packing fraction upon the particle–plate contact area (Ramakrishnan and Pillai, 1980).

Field-induced solidification phenomena in magnetic and electric field–controlled moving granular beds do not really anticipate the unique properties of the magnetostabilized bed (MSB) (Rosensweig, 1979b). The MSB uses a highly uniform DC magnetic field usually oriented parallel to the flow direction. Unlike the electrofluidized bed, the MSB expands uniformly as the flow rate increases. Bubbling is either greatly reduced or eliminated altogether, while the minimum fluidization condition is left unchanged. MSB behavior differs from the DC electrofluidized bed because the essential particle electromechanics stem from the multipolar interactions of particles not in intimate contact. Particle packing in the MSB has been investigated by examining matrix sections cut from a bed that was fixed in a polymer matrix (Rosensweig et al., 1981). Interestingly, the method reveals only a modest degree of anisotropy, with particles loosely oriented in lines parallel to the field.[7]

Under conditions not too different from those employed to achieve the magnetostabilized bed, slugs of magnetizable particles in pipes also can be made to exhibit irreversible *plastic* yield phenomena (von Guggenburg et al., 1986).

7.7 Closing prospect

Looking back over the seven chapters of this book, the reader inevitably will be struck by the number and the amazing diversity of technological applications where electric or magnetic fields are employed to control particles. In fact, such engineered particulate systems grow steadily more important with each passing year. The author has striven to relate his subject – particle electromechanics – to these industrial and commercial applications and to scientific endeavors. Chapters 1 through 5 have focused on fundamental electromechanical mechanisms –

---

[7] When an AC electric field is applied to beds consisting of particles with high dielectric constant, stabilized fluidization is also achieved (Zahn and Rhee, 1984). In general, the behavior of such *electrostabilized beds* is comparable to the magnetostabilized bed.

forces, rotational, and alignment torques – that are now being exploited in the control and manipulation of single particles. Chapters 6 and 7 have investigated the strong particle interactions that lead to chain formation and cause the unique electromechanical behavior of particulate layers and suspensions. In particular, Sections 7.5 and 7.6 have identified many important or potentially important technologies where fields are used to control the motion or orientation of chains, layers, or beds of particles in the size range from 1 to 1000 μm. All these examples should convince us of the emerging economic and technological importance of particle electromechanics as a scientific and engineering discipline.

Throughout this book has run a common thread – use of the method of effective multipoles for investigating electromechanical interactions of particles subjected to uniform, nonuniform, and rotating fields. The first application of the method has been in calculation of forces of electrical origin, principally the dielectrophoretic (DEP) and magnetophoretic (MAP) forces on single particles. It has then been extended to determination of rotational and orientational torques, also on single particles. Despite the complications of layered particles, the introduction of relaxation mechanisms and ohmic loss, and even material anisotropy and nonlinearity, this method is found to serve us well as a systematic way to treat the electromechanics of particles. Building on the results obtained with single particles, the method of effective multipoles has been extended and generalized to the far more complicated cases of chains and layers, as exemplified by such examples as the magnetic brush xerographic copy engine, electrorheological fluids, and magnetically coupled granular beds. Though only a modest start has been made at it here, the method of effective multipoles should provide a valuable tool for developing predictive models for the electromechanics of powders, slurries, and granular beds.

In any densely packed powder, granular bed, or particle–liquid suspension, dielectric (or magnetic) interactions among particles are very complex because large numbers of neighbors become involved. The chain models developed in Chapters 6 and 7 of this book, while serving as a starting point, will have to be replaced by more complete representations of the strong field-induced coupling. What is required next is a marriage of the disciplines of particle electromechanics and powder mechanics to form a new discipline – powder electromechanics. While no attempt to achieve this difficult union has been made in the present book, the author hopes that the systematic framework for treating particle electromechanics offered here will ultimately help to bring it about.

Appendix A

Analogies between electrostatic, conduction, and magnetostatic problems

A. Introduction

The purpose of this appendix is to summarize a set of important relationships comprising a very useful analogy in the electromechanics of particles. Though this analogy has far broader application in the engineering sciences, we will identify here only three distinct cases: (i) electrostatic problems with insulating linear dielectrics containing no free charge, (ii) electrostatic problems with linear ohmic conductors and DC electric fields, and (iii) magnetostatic problems with nonconducting linear magnetizable media.[1] The analogy, which is applicable only in the case when linear constitutive relationships exist for materials, may be used to apply the results obtained for one type of problem to another type by a simple set of variable and parameter substitutions without having to resort to further analysis. We first summarize the basic physical laws and assumptions applicable for each of the analogous physical problems.

B. Basic relationships

We reveal the analogy by writing out the basic field relationships and then comparing them to find canonical similarities and, subsequently, to identify the analogous quantities. The first basic field relationship involves the curl of the vector field intensity. For the dielectric and conductive cases, the curl-free condition on $\bar{E}$ essentially defines the electrostatic assumption. For the magnetostatic case, the curl-free condition on $\bar{H}$ means that there are no current sources in the solution region. These relationships are written down below in column format to facilitate comparison among the three different types of problems.

Dielectric case	Conductive case	Magnetostatic case
$\nabla \times \bar{E} = 0$	$\nabla \times \bar{E} = 0$	$\nabla \times \bar{H} = 0$

[1] Dukhin (1971) recognizes this analogy and presents it in terms of what he calls the "generalized conductivity."

A second set of analogous relationships arises from setting the divergence of a vector flux quantity to zero. Each one of these conditions signifies a distinct physical assumption. Specifying the electric flux or displacement vector $\bar{D}$ to be divergence-free for the dielectric case means that there is no volume charge. For the conductive case, a divergence-free current density $\bar{J}$ means that the volume charge density is constant at all points and, furthermore, that the problem is restricted to steady-state DC conditions, i.e., $\partial/\partial t = 0$. Setting the divergence of the magnetic flux $\bar{B}$ to zero recognizes that there are no magnetic monopoles.

$$\nabla \bullet \bar{D} = 0 \qquad\qquad \nabla \bullet \bar{J} = 0 \qquad\qquad \nabla \bullet \bar{B} = 0$$

The next set of relationships are linear constitutive laws that relate the respective flux and field intensity vectors. The quantities ε, σ, and μ are, respectively, dielectric permittivity (F/m), electrical conductivity (S/m), and magnetic permeability (H/m).

$$\bar{D} = \varepsilon\bar{E} \qquad\qquad \bar{J} - \sigma\bar{E} \qquad\qquad \bar{B} = \mu\bar{H}$$

Finally, subsidiary relationships exist that are based upon the set of curl-free field intensity conditions stated above. These conditions involve the definition of various scalar potential functions.

$$\bar{E} \equiv -\nabla\Phi \qquad\qquad \bar{E} \equiv -\nabla\Phi \qquad\qquad \bar{H} \equiv -\nabla\Psi$$

The electrostatic scalar potential Φ is defined for both the dielectric and conduction problems, while Ψ is known as the magnetostatic scalar potential.

C. Boundary conditions

Though the boundary conditions are based upon the above fundamental relationships, still it is worthwhile stating them here to help explain the nature of the analogy. The first boundary condition, due to the curl-free conditions imposed upon $\bar{E}$ and $\bar{H}$, requires that the tangential components of these fields be continuous across the boundary between two regions #1 and #2, where $\hat{n}$, the unit normal vector at the interface, points into material #1.

$$\hat{n} \times (\bar{E}_1 - \bar{E}_2) = 0 \qquad\qquad \hat{n} \times (\bar{E}_1 - \bar{E}_2) = 0 \qquad\qquad \hat{n} \times (\bar{H}_1 - \bar{H}_2) = 0$$

An entirely equivalent statement in terms of the scalar potentials is:

$$\Phi_1 = \Phi_2 \qquad\qquad \Phi_1 = \Phi_2 \qquad\qquad \Psi_1 = \Psi_2$$

Another boundary condition derived from the divergence condition describes the continuity of the normal component of the flux.

$$\hat{n} \bullet (\bar{D}_1 - \bar{D}_2) = 0 \qquad\qquad \hat{n} \bullet (\bar{J}_1 - \bar{J}_2) = 0 \qquad\qquad \hat{n} \bullet (\bar{B}_1 - \bar{B}_2) = 0$$

D. Analogous quantities

Direct comparison of the relationships set out side by side above reveals the analogy, and the analogous variables are summarized in Table A.1, along with their proper SI units in brackets. Note that the Clausius–Mossotti function K in this table is applicable only to spherical particles.

E. Force and torque expressions

Table A.2 provides expressions for the effective dipole moment (for spherical particles) plus the nonuniform field force and electrical torque on these moments. Certain minor distinctions noted among these expressions exist because of the absence of symmetry between the definitions used in formulations of Maxwell's equations and the Lorentz force law.

F. Discussion

Chapters 6 and 7 will convince the reader that from the standpoint of basic particle interactions, the distinctions between magnetic and electric field particle interactions are more a matter of algebraic symbols and units than anything fundamental. Even in the presence of nonlinear behavior, the essential features of the analogy seem to remain. Furthermore, striking resemblances exist between certain electric and magnetic particle systems in the realm of engineered particulate technologies. For example, compare electrofluidized and electrospouted beds to their magnetofluidized and magnetospouted bed counterparts. Another close parallel exists between the magnetic powder coupling and the electropacking of a precipitated dust layer in an electrostatic precipitator. Despite very distinct differences in the electrostatic and magnetostatic origins of the interparticle forces, the predicted and observed electromechanical phenomenology is common to all.

Table A.1 *Summary of some important analogous variables and parameters for dielectric, DC conductive, and magnetostatic problems, all having linear constitutive laws*

Quantity	Dielectric case	Conductive case	Magnetostatic case
Field intensity	$\bar{E}\,[\text{V/m}]$	$\bar{E}\,[\text{V/m}]$	$\bar{H}\,[\text{A/m}]$
Flux density	$\bar{D}\,[\text{C/m}^2]$	$\bar{J}\,[\text{A/m}^2]$	$\bar{B}\,[\text{T}]$
Polarization	$\bar{P}\,[\text{C/m}^2]$	$\bar{P}\,[\text{C/m}^2]$	$\bar{M}\,[\text{A/m}]$
Scalar potential	$\Phi\,[\text{V}]$	$\Phi\,[\text{V}]$	$\Psi\,[\text{A}]$
Linear law coefficient	$\varepsilon\,[\text{F/m}]$	$\sigma\,[\text{S/m}]$	$\mu\,[\text{H/m}]$
Clausius–Mossotti function (K)	$\dfrac{\varepsilon_2 - \varepsilon_1}{\varepsilon_2 + 2\varepsilon_1}$	$\dfrac{\sigma_2 - \sigma_1}{\sigma_2 + 2\sigma_1}$	$\dfrac{\mu_2 - \mu_1}{\mu_2 + 2\mu_1}$

Table A.2 *Summary of the effective moment expressions, plus the gradient force and torque expressions for dielectric, DC conductive, and magnetostatic spheres, assuming a linear constitutive law in each case*

Quantity	Dielectric case	Conductive case	Magnetostatic case
Effective Dipole moment	$4\pi\varepsilon_1 R^3\left(\dfrac{\varepsilon_2 - \varepsilon_1}{\varepsilon_2 + 2\varepsilon_1}\right)\bar{E}$	$4\pi\varepsilon_1 R^3\left(\dfrac{\sigma_2 - \sigma_1}{\sigma_2 + 2\sigma_1}\right)\bar{E}$	$4\pi R^3\left(\dfrac{\mu_2 - \mu_1}{\mu_2 + 2\mu_1}\right)\bar{H}$
Force on dipole	$(\bar{p}_{\text{eff}} \cdot \nabla)\bar{E}$	$(\bar{p}_{\text{eff}} \cdot \nabla)\bar{E}$	$\mu_1(\bar{m}_{\text{eff}} \cdot \nabla)\bar{H}$
Torque on dipole	$\bar{p}_{\text{eff}} \times \bar{E}$	$\bar{p}_{\text{eff}} \times \bar{E}$	$\mu_1 \bar{m}_{\text{eff}} \times \bar{H}$

Appendix B

Review of linear multipoles

A. *Introduction*

This appendix summarizes the application of linear multipoles in expansions of the electrostatic potential for axisymmetric distributions of electric charge. Included are expressions for the forces exerted on linear multipoles by nonuniform axisymmetric electric fields. Many of the important results in particle electromechanics can be modeled successfully using these force expressions.

B. *Field due to a finite dipole*

Consider the electrostatic potential Φ due to a *finite dipole*, that is, two point charges $+q$ and $-q$ immersed in a linear dielectric of permittivity ε_1 and located on the z axis near the origin at $z = \pm d/2$, as shown in Figure B.1. This charge distribution is axisymmetric so that $\Phi = \Phi(r,\theta)$, where r is the radial coordinate and θ is the polar angle in spherical coordinates. In the limit where $d \to 0$ and $q \to \infty$

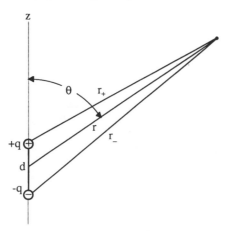

Fig. B.1 Small physical dipole in dielectric fluid of permeability ε_1 aligned with z axis showing r_+ and r_- defined at an arbitrary point (r,θ).

Table B.1 *The first several Legendre polynomial terms used for expressing the electrostatic potential due to the linear multipoles, or any axisymmetric charge distribution*

n	$P_n(\cos\theta)$
0	1
1	$\cos\theta$
2	$[3\cos^2\theta - 1]/2$
3	$[5\cos^3\theta - 3\cos\theta]/2$

such that the qd product remains finite, the point dipole results. Here, however, we take into account the finite spacing of the two finite charges and invoke superposition to express the net electrostatic potential Φ.

$$\Phi(r,\theta) = \frac{q}{4\pi\varepsilon_1 r_+} - \frac{q}{4\pi\varepsilon_1 r_-} \tag{B.1}$$

From simple geometric considerations, we may show that r_+ and r_-, defined in Figure B.1, are related to d and (r,θ) as follows:

$$\left(\frac{r}{r_\pm}\right) = \left[1 + \left(\frac{d}{2r}\right)^2 \mp \frac{d}{r}\cos\theta\right]^{-1/2} \tag{B.2}$$

Equation (B.2) is now expanded using the familiar Maclaurin series

$$(1+x)^{-1/2} = 1 - \frac{x}{2} + \frac{3x^2}{8} - \frac{5x^3}{16} + \cdots \tag{B.3}$$

Employing Equation (B.3) with (B.2), we obtain

$$\left(\frac{r}{r_\pm}\right) = P_0 \pm \left(\frac{d}{2r}\right)P_1 + \left(\frac{d}{2r}\right)^2 P_2 \pm \left(\frac{d}{2r}\right)^3 P_3 + \cdots \tag{B.4}$$

where $P_1(\cos\theta)$, $P_2(\cos\theta)$, ..., are the Legendre polynomials. Table B.1 provides expressions for the first several of these polynomials.

Combining Equations (B.4) and (B.1) yields an expression for the electrostatic potential of the small physical dipole shown in Figure B.1. The result is

$$\Phi(r,\theta) = \frac{qd P_1(\cos\theta)}{4\pi\varepsilon_1 r^2} + \frac{qd^3 P_3(\cos\theta)}{16\pi\varepsilon_1 r^4} + \cdots \tag{B.5}$$

where the dipole term has the same form as Φ_{dipole} defined by Equation (2.7) and

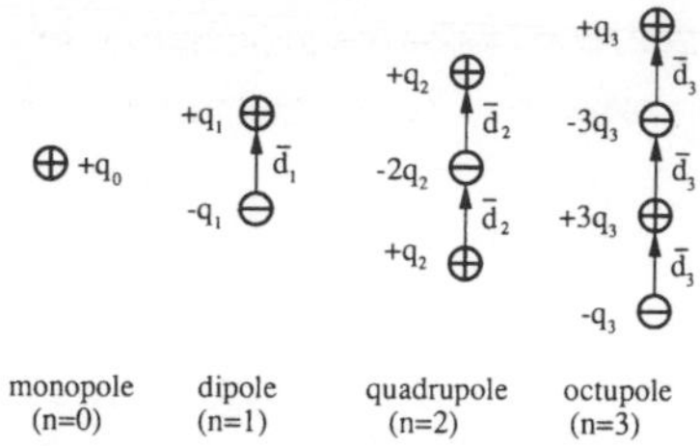

Fig. B.2 First several linear multipoles ($n = 0$, 1, 2, and 3) as constructed from point charges of alternating signs.

the next nonzero term ($n = 3$) is an octupolar correction. Note that all additional, higher-order terms are of odd order, that is, $n = 5, 7, \ldots$, for the finite dipole.

C. The linear multipoles

Figure B.2 shows how to construct all the linear multipoles using evenly spaced point charges of appropriate magnitude and sign. Note that these magnitudes correspond to the terms of the binomial expansion. Using the superposition principle and Equation (B.4), expressions for the electrostatic potential of each multipole may be derived. For a multipole of order n, the potential Φ_n is

$$\Phi_n = \frac{p^{(n)}}{4\pi\varepsilon_1 r^{n+1}} P_n(\cos\theta) \tag{B.6}$$

and the general expression for the multipolar moment of order n, as defined by Figure B.2, is

$$p^{(n)} = n! q_n d_n^n \tag{B.7}$$

Here, q_n and d_n represent, respectively, the unit charge and the separation of the point charges constituting the nth linear multipolar distribution.

We can show that the electrostatic potential of any axisymmetric charge distribution may be expressed as a summation of linear multipolar terms as given by Equation (B.6) (Becker, 1964, Sec. 24). An alternative derivation of Equation (B.6) employs the method of separation of variables on Laplace's equation in spherical coordinates for the case of axisymmetry.

As a special case, consider an array of N point charges ($q_1, q_2, \ldots, q_N$) distributed along a line. The nth linear multipolar moment of this array is

$$p^{(n)} = \sum_{i=1}^{N} q_i d_i^n \tag{B.8}$$

Equation (B.8) reveals that, in general, the moments of a charge distribution are functions of the choice of origin. For example, the dipole moment becomes independent of the choice of origin only when the net charge is zero, that is,

$$\sum_{i=1}^{N} q_i = 0.$$

D. Forces on linear multipoles

To model the mutual electromechanical interactions of polarized particles in electric fields using multipoles, we need expressions for the forces exerted by an arbitrary electric field upon the nth multipole. The case of the simple dipole in a nonuniform electric field, as depicted in Figure 2.1a, already has been considered in Section 2.1B. A similar analysis applied to the linear quadrupole illustrated in Figure B.3 yields

$$\bar{F}_{\text{quadrupole}} = q_2(\bar{d}_2 \cdot \nabla)^2 \bar{E}_0 \tag{B.9}$$

By suitable generalization, we may obtain an expression for the force exerted on the nth order linear multipole by an electric field.

$$\bar{F}_n = q_n(\bar{d}_n \cdot \nabla)^n \bar{E}_0 \tag{B.10}$$

For the special case of multipoles located on and aligned with the axis in an axisymmetric electric field, the radial component of the force must be zero. The purely z-directed force then takes the form

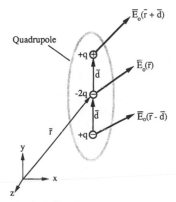

Fig. B.3 Free body diagram for determining net electrostatic force on a linear quadrupole.

Table B.2 *Mathematical expressions for the moment $p^{(n)}$, electrostatic potential Φ_n, and z-directed force F_n of first four linear multipoles ($n = 0, 1, 2,$ and 3), plus the general multipole of order n*

Multipolar term	Moment: $p^{(n)}$	Potential: Φ_n	Force: $\bar{F}_n$
Monopole: $n = 0$	$p^{(0)} = q_0$	$\dfrac{q_0}{4\pi\varepsilon_1 r}$	$q_0 \bar{E}$
Dipole: $n = 1$	$p^{(1)} = q_1 d_1$	$\dfrac{q_1 d_1}{4\pi\varepsilon_1 r^2}(\cos\theta)$	$(q_1 \bar{d}_1 \cdot \nabla)\bar{E}$
Quadrupole: $n = 2$	$p^{(2)} = 2q_2 d_2^2$	$\dfrac{2q_2 d_2^2}{4\pi\varepsilon_1 r^3}\left(\dfrac{3\cos^2\theta - 1}{2}\right)$	$q_2(\bar{d}_2 \cdot \nabla)^2 \bar{E}$
Octupole: $n = 3$	$p^{(3)} = 6q_3 d_3^3$	$\dfrac{6q_3 d_3^3}{4\pi\varepsilon_1 r^4}\left(\dfrac{5\cos^3\theta - 3\cos\theta}{2}\right)$	$q_3(\bar{d}_3 \cdot \nabla)^3 \bar{E}$
General nth-order multipole	$p^{(n)} = n! q_n d_n^n$	$\Phi_n = \dfrac{p^{(n)}}{4\pi\varepsilon_1 r^{n+1}} P_n(\cos\theta)$	$\bar{F}_n = q_n(\bar{d}_n \cdot \nabla)^n \bar{E}$

$$\bar{F}_n = \frac{p^{(n)}}{n!} \frac{\partial^n E_z}{\partial z^n}\hat{z} \tag{B.11}$$

with the expression for the moment $p^{(n)}$ provided by Equation (B.7).

E. Summary

This appendix sets forth results for the electrostatic potential due to and the force exerted by an axisymmetric electric field upon linear multipoles. A more general effective multipolar theory, outlined in Appendix F, uses dyadic tensor representation for the moments. Nevertheless, linear multipoles serve adequately to explain the essential aspects of dielectrophoresis and particle interactions in most practical situations. Table B.2 gives the electrostatic potential and the net force expressions for the monopole, dipole, quadrupole, and octupole ($n = 0$ through 3). For convenience, expressions for the general case of the nth-order multipole are also included.

Appendix C
Models for layered spherical particles

A. Introduction

This appendix summarizes a simple-to-use method for modeling layered spherical particles. According to this method, a homogeneous sphere with permittivity ε_2' is substituted for the original particle. For a uniform or nearly uniform electric field, only the dipole is induced and only a single effective value for ε_2' is identified. On the other hand, in a nonuniform electric field, different effective permittivity values are required for each multipolar term and no unique value for ε_2' may be identified.

B. Lossless spherical shell in uniform field

We present the important formal details of the approach by considering first a lossless dielectric sphere covered with a single layer of uniform thickness. Figure C.1 shows a layered lossless dielectric sphere subjected to a uniform electric field $\bar{E}_0 = E_0 \hat{z}$. The concentric outer shell and inner core have permittivities ε_2 and ε_3, respectively, while the particle radius is R_1 and the core radius is R_2. For the present, we assume that there is no free (unpaired) electric charge anywhere. To solve this boundary value problem, we assume solutions to Laplace's equation for the electrostatic potential Φ in the three regions.

$$\Phi_1 = \left(-E_0 r + \frac{A}{r^2}\right)\cos\theta, \ r > R_1 \tag{C.1a}$$

$$\Phi_2 = \left(-Br + \frac{C}{r^2}\right)\cos\theta, \ R_1 > r > R_2 \tag{C.1b}$$

$$\Phi_3 = -Dr\cos\theta, \ r < R_2 \tag{C.1c}$$

The boundary conditions at the two dielectric interfaces are

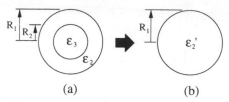

Fig. C.1 Layered spherical shell (a) and equivalent homogeneous sphere (b) with apparent homogeneous dielectric permittivity ε_2' defined by Equation (C.4).

$$\Phi_1 = \Phi_2 \ \text{and} \ \varepsilon_1 \frac{\partial \Phi_1}{\partial r} = \varepsilon_2 \frac{\partial \Phi_2}{\partial r}, \ \text{at} \ r = R_1 \tag{C.2a}$$

$$\Phi_2 = \Phi_3 \ \text{and} \ \varepsilon_2 \frac{\partial \Phi_2}{\partial r} = \varepsilon_3 \frac{\partial \Phi_3}{\partial r}, \ \text{at} \ r = R_2 \tag{C.2b}$$

According to Equation (2.11), only the constant A is required for determination of the effective dipole moment p_{eff}; nevertheless, for the sake of completeness, all coefficients are given here. Combining the assumed solutions with the boundary conditions, we obtain

$$A = \frac{\varepsilon_2' - \varepsilon_1}{\varepsilon_2' + 2\varepsilon_1} R_1^3 E_0 \tag{C.3a}$$

$$B = -\frac{3\varepsilon_1 a^3}{(\varepsilon_2' + 2\varepsilon_1)(a^3 - K)} E_0 \tag{C.3b}$$

$$C = \frac{3\varepsilon_1 K R_1^3}{(\varepsilon_2' + 2\varepsilon_1)(a^3 - K)} E_0 \tag{C.3c}$$

$$D = -\frac{3\varepsilon_1 (1 - K) a^3}{(\varepsilon_2' + 2\varepsilon_1)(a^3 - K)} E_0 \tag{C.3d}$$

where $a = R_1/R_2$. An effective homogeneous dielectric permittivity value ε_2' replaces ε_2 in the Clausius–Mossotti factor, i.e., $K = (\varepsilon_2' - \varepsilon_1)/(\varepsilon_2' + 2\varepsilon_1)$.

$$\varepsilon_2' = \varepsilon_2 \left\{ \frac{a^3 + 2\left(\dfrac{\varepsilon_3 - \varepsilon_2}{\varepsilon_3 + 2\varepsilon_2}\right)}{a^3 - \left(\dfrac{\varepsilon_3 - \varepsilon_2}{\varepsilon_3 + 2\varepsilon_2}\right)} \right\} \tag{C.4}$$

Introduction of ε_2' simplifies the algebraic expressions for the coefficients and, in particular, puts A into the familiar form of Equation (2.10). Evidently, the electrostatic potential solution *outside* the layered sphere is indistinguishable from that of the equivalent homogeneous particle with radius R_1 and permittivity ε'_2 shown in Figure C.1b. Equation (C.4) is identical to Maxwell's classic mixture formula (Maxwell, 1954, art. 314) for the effective permittivity of a mixture of noninteracting spherical particles of permittivity ε_2 in a fluid medium of permittivity ε_1 with volume fraction (particle volume–mixture volume) equal to a^3 (Pauly and Schwan, 1959).

This method is easily applied to multilayered shells using the approach illustrated in Figure C.2. By starting at the innermost layer and repeatedly applying Equation (C.4), we eventually arrive at the desired expression for ε'_2, the permittivity of the equivalent homogeneous sphere.

C. Spherical shell in uniform field with ohmic loss

Generalization of Equation (C.4) to account for the important case of dielectric particles with loss is not a difficult task. Imagine the layered dielectric sphere of Figure C.1a with permittivities and conductivities (ε_1,σ_1), (ε_2,σ_2), and (ε_3,σ_3) associated with the external medium $(r > R_1)$, the shell $(R_1 > r > R_2)$ and the core $(r < R_2)$, respectively. The externally imposed, spatially uniform AC electric field is $\bar{E}(t) = \mathrm{Re}[E_0 \hat{z}\exp(j\omega t)]$ and ω is the radian frequency. The general solution form of Equations (C.1a,b,c) is unchanged, except that the coefficients A, B, C, and D are now complex quantities. We modify the boundary conditions, Equations (C.2a,b), to account for Maxwell–Wagner surface charge by replacement of ε_1, ε_2, and ε_3 with their complex equivalents, that is, $\underline{\varepsilon}_1, \underline{\varepsilon}_2$, and $\underline{\varepsilon}_3$, where $\underline{\varepsilon}_1 = \varepsilon_1 + \sigma_1/j\omega$, etc.

Using the above transformations, an expression for the complex coefficient $\underline{A}$ results.

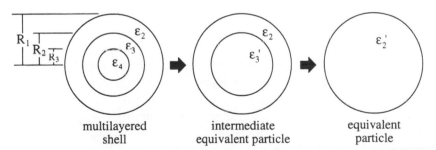

| multilayered shell | intermediate equivalent particle | equivalent particle |

Fig. C.2 Repeated application of Equation (C.4) may be used to replace a multilayered dielectric spherical shell by an equivalent homogeneous sphere of radius R_1 and permittivity ε'_2.

$$\underline{A} = \frac{\varepsilon_2' - \varepsilon_1}{\varepsilon_2' + \varepsilon_1} R_1^3 E_0 \tag{C.5}$$

where

$$\varepsilon_2' = \varepsilon_2 \left\{ \frac{a^3 + 2\left(\dfrac{\varepsilon_3 - \varepsilon_2}{\varepsilon_3 + 2\varepsilon_2}\right)}{a^3 - \left(\dfrac{\varepsilon_3 - \varepsilon_2}{\varepsilon_3 + 2\varepsilon_2}\right)} \right\} \tag{C.6}$$

Note that the effective values of the scalar permittivity and conductivity now depend upon excitation frequency ω.

$$\varepsilon_2'(\omega) = \text{Re}(\varepsilon_2') \quad \text{and} \quad \sigma_2'(\omega) = -\omega \text{Im}(\varepsilon_2') \tag{C.7}$$

This frequency dependence, a consequence of Maxwell–Wagner polarization at the $r = R_2$ interface, vanishes only under the special condition that $\varepsilon_2/\sigma_2 = \varepsilon_3/\sigma_3$. When this condition is met, no time-varying free charge accumulates at the $r = R_2$ interface. Note that this condition does not influence possible free charge accumulation at the outer surface of the particle, $r = R_1$.

   Based on Equation (2.11), the complex effective dipole moment vector must be $\bar{p}_{\text{eff}} = 4\pi\varepsilon_1 \underline{A}\hat{z}$ and so the instantaneous values of this sinusoidally varying moment can be written out using Equation (C.5).

$$\bar{p}_{\text{eff}}(t) = 4\pi\varepsilon_1 R_1^3 \text{Re}\left[\frac{\varepsilon_2' - \varepsilon_1}{\varepsilon_2' + 2\varepsilon_1} E_0 \hat{z}\exp(j\omega t)\right] \tag{C.8}$$

See Section 2.3A for a discussion of why the factor $4\pi\varepsilon_1$ (*not* $4\pi\varepsilon_1$) appears in Equation (C.8). The time-dependent expression for effective moment is important in force calculations using the effective moment method.

D. Thin surface layers

The case of very thin surface coatings and layers is quite important in modeling many types of real particles, such as biological cells and colloidal particles. Consider the lossy dielectric sphere with a surface layer of thickness $\Delta = R_1 - R_2$ $<< R$ shown in Figure C.3. We may identify two important general limits for such thin surface layers: (i) the *series admittance limit*, where a finite electrostatic potential can be supported across the layer; and (ii) the *shunt admittance limit*, where finite current can flow in the layer tangential to the particle's surface. Though it is possible to extract these limits directly from Equation (C.6),

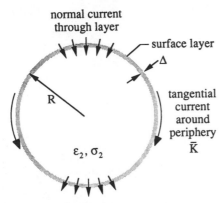

Fig. C.3 Particle with very thin outer layer of thickness Δ, showing the pathways for electric current and displacement flux through the layer (series flow) and around the periphery (shunt flow).

more physical insight is gained from separate consideration of the boundary value problems for the two limiting cases.

D.1 Series admittance model

Consider a spherical particle with radius R, permittivity ε_2, and conductivity σ_2, which is covered by a uniform layer of thickness $\Delta \ll R$, permittivity ε_m, and ohmic conductivity $\sigma_m \ll \sigma_2$ supporting a finite potential drop. Such layers are conveniently modeled in terms of surface capacitance c_m ($= \varepsilon_m/\Delta$, in F/m^2) and surface transconductance g_m ($= \sigma_m/\Delta$, in S/m^2). We can approach this boundary value problem by assuming electrostatic potential solutions of the same form as Equation (2.8a,b). Here, it is convenient to express the revised current continuity conditions in terms of the radial components of the electric field $E_r = -\partial\Phi/\partial r$ and the electrostatic potential Φ.

$$\underline{\varepsilon}_1\underline{E}_{r1} = \underline{\varepsilon}_2\underline{E}_{r2}, \quad \text{at } r = R \tag{C.9a}$$

and

$$(j\omega c_m + g_m)(\underline{\Phi}_1 - \underline{\Phi}_2) = -j\omega\underline{\varepsilon}_1\underline{E}_{r1} \quad \text{at } r = R \tag{C.9b}$$

Equation (C.9b) specifically accounts for the finite voltage drop across the layer in terms of capacitance and transconductance per unit area. Solution for the coefficient $\underline{A}$ yields an expression identical to Equation (C.5) in its form but with a new expression for effective permittivity.

$$\underline{\varepsilon}_2' = \frac{\underline{c}_m R \underline{\varepsilon}_2}{\underline{c}_m R + \underline{\varepsilon}_2} \tag{C.10}$$

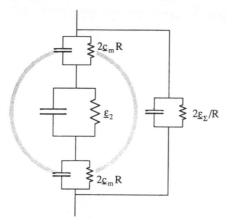

Fig. C.4 Equivalent *RC* circuit representing series and shunt paths for electric current and flux. Refer to Equations (C.10) and (C.15) in the text.

For convenience, we define the complex capacitance per unit area $\underline{c}_m = c_m + g_m/j\omega$. Then, Equation (C.10) takes the form of the admittance of two series-connected circuit elements ε_2 and $\underline{c}_m R$, as illustrated in Figure C.4.

The series admittance model is a convenient representation for certain biological cells because the membrane structure common to all cells, consisting of a lipid protein bi-layer, is very thin ($\Delta \approx 100$ Å $= 0.01$ µm) compared to overall cell dimensions. The membrane capacitance c_m is ordinarily ~1 µF/cm^2, while the transmembrane conductance of healthy cells is small ($g_m < 1$ mS/cm^2) and has negligible effect except at very low frequencies. The cell interior or cytoplasm is an aqueous electrolyte with $\varepsilon_2 \sim 80\varepsilon_0$ and $\sigma_2 \sim 0.5$ S/m. Because the membrane supports a potential drop over a wide range of electric field frequencies, it serves as an example of a surface layer accurately modeled by series admittance. Transconductance g_m becomes significant when the cell is senescent or if the membrane is altered by chemical insult or strong electric field pulses.

D.2 Shunt admittance model

If the surface layer on a particle supports the flow of ohmic or displacement current tangent to the surface, then some of the current incident normal to the particle will be shunted around the periphery without flowing through the interior. In this situation, a shunt admittance model is appropriate for the particle's effective complex permittivity. A common example where the shunt admittance model is appropriate is an insulating particle with a very thin film of adsorbed water.

Consider the spherical particle of Figure C.3 with radius R, permittivity ε_2, and conductivity σ_2, and covered by a layer of thickness $\Delta \ll R$. If this thin layer has sufficiently high conductivity, it will support finite tangential as well as normal current. We approach the boundary value problem by again starting with the electrostatic potential solutions of Equations (2.8a,b). The electrostatic potential is continuous across the boundary,

$$\underline{\Phi}_1 = \underline{\Phi}_2, \text{ at } r = R \tag{C.11}$$

The boundary condition enforcing charge continuity balances normal and tangential components of the electric current with free surface charge buildup.

$$j\omega[\underline{\varepsilon}_1\underline{E}_{r1} - \underline{\varepsilon}_2\underline{E}_{r2}] + \nabla_\Sigma \cdot \overline{\underline{K}} = 0, \text{ at } r = R \tag{C.12}$$

Here, $E_r = -\partial\Phi/\partial r$ and $E_\theta = -r^{-1}\partial\Phi/\partial\theta$ are the radial and tangential components of the electric field, respectively. $\overline{\underline{K}}$ is surface current density and ∇_Σ is the surface del operator, so that

$$\nabla_\Sigma \cdot \overline{\underline{K}}\big|_{r=R} = \hat{\theta}\frac{1}{R\sin\theta}\frac{\partial[\sin\theta\underline{K}_\theta]}{\partial\theta} + \hat{\phi}\frac{1}{R\sin\theta}\frac{\partial[\underline{K}_\phi]}{\partial\phi} \tag{C.13}$$

Now, assume that the surface current $\overline{\underline{K}}$ is proportional to the tangential electric field.

$$\overline{\underline{K}} = j\omega\underline{\varepsilon}_\Sigma\overline{\underline{E}}_{tan} = j\omega\underline{\varepsilon}_\Sigma\underline{E}_\theta(r=R,\theta)\,\hat{\theta} \tag{C.14}$$

where $\underline{\varepsilon}_\Sigma = \varepsilon_\Sigma + \sigma_\Sigma/j\omega$. Here, ε_Σ, interpreted as surface permittivity (in farads), accounts for any electric field–induced out-of-phase motion of the ionic charge layer typically surrounding a biological cell or colloidal particle in aqueous suspension (Schwarz, 1962), while σ_Σ is the more familiar ohmic surface conductivity (in siemens). Solution for $\underline{A}$ again yields Equation (C.5), but with still another form for the effective complex permittivity, namely

$$\underline{\varepsilon}'_2 = \underline{\varepsilon}_2 + \frac{2\underline{\varepsilon}_\Sigma}{R} \tag{C.15}$$

If $\underline{\varepsilon}_2$ and $2\underline{\varepsilon}_\Sigma/R$ are considered as admittance terms, then Equation (C.15) can be interpreted as the total admittance of two shunt-connected circuit elements. Refer again to Figure C.4.

D.3 Model for particle with surface resistance

The special case of a particle with a resistive layer of conductivity σ_Σ is very important because many different types of small particles are treated with or become covered by surface coatings that increase the overall effective conductivity. Consider the limit of Equation (C.15) where $\varepsilon_2 \gg \varepsilon_\Sigma/R$. Then, we have

$$\varepsilon_2' \to \varepsilon_2 + \frac{\sigma_2 + \dfrac{2\sigma_\Sigma}{R}}{j\omega} \tag{C.16}$$

This expression represents the fairly common situation of a poorly conducting particle with a semiconducting outer layer. Surface conductance increases the effective (DC) conductance of the particle by the radius-dependent factor $2\sigma_\Sigma/R$.

E. Lossless spherical shell in nonuniform field

Figure C.5 shows a concentrically layered spherical particle of outer radius R_1 immersed in a uniform dielectric fluid of permittivity ε_1 and located a distance ζ from the point source charge q. We seek expressions for the multipolar moments $p^{(n)}$ induced in this heterogeneous particle by the field. Consistent with the approach taken in Section 2.2E, we choose the following assumed solutions for the electrostatic potential in the three regions.

$$\Phi_1 = \frac{q}{4\pi\varepsilon_1\zeta} \sum_{n=0}^{\infty} \left(\frac{r}{\zeta}\right)^n P_n(\cos\theta) + \sum_{n=0}^{\infty} \frac{A_n P_n(\cos\theta)}{r^{n+1}}, \ r \geq R_1 \tag{C.17a}$$

$$\Phi_2 = \sum_{n=0}^{\infty} B_n r^n P_n(\cos\theta) + \sum_{n=0}^{\infty} \frac{C_n P_n(\cos\theta)}{r^{n+1}}, \ R_1 \geq r \geq R_2 \tag{C.17b}$$

$$\Phi_3 = \sum_{n=0}^{\infty} D_n r^n P_n(\cos\theta), \ R_2 \geq r \tag{C.17c}$$

In the above, we have used Equation (2.18) to replace $1/r_q$ with an expansion of Legendre polynomial terms. The boundary conditions are unchanged from

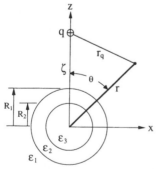

Fig. C.5 Spherical dielectric shell in the field of a point charge q.

Equations (C.2a,b) above. To identify the effective multipolar moments $p^{(n)}$, it is necessary to solve only for the coefficients A_n, though for the sake of completeness, all coefficients are provided below.[1]

$$A_n = -\left\{\frac{n\,[(\varepsilon_2)'_n - \varepsilon_1]}{n(\varepsilon_2)'_n + (n+1)\varepsilon_1}\right\}\frac{R_1^{2n+1}\,q}{4\pi\varepsilon_1\zeta^{n+1}} \tag{C.18a}$$

$$B_n = \left\{\frac{(2n+1)\varepsilon_1 R_2^{-(2n+1)}}{[n(\varepsilon_2)'_n + (n+1)\varepsilon_1][a^{2n+1} - nK^{(n)}]}\right\}\frac{R_1^{2n+1}\,q}{4\pi\varepsilon_1\zeta^{n+1}} \tag{C.18b}$$

$$C_n = -\left\{\frac{n(2n+1)\varepsilon_1 K^{(n)}}{[n(\varepsilon_2)'_n + (n+1)\varepsilon_1][a^{2n+1} - nK^{(n)}]}\right\}\frac{R_1^{2n+1}\,q}{4\pi\varepsilon_1\zeta^{n+1}} \tag{C.18c}$$

$$D_n = \left\{\frac{(2n+1)\varepsilon_1[1 - nK^{(n)}]R_2^{-(2n+1)}}{[n(\varepsilon_2)'_n + (n+1)\varepsilon_1][a^{2n+1} - nK^{(n)}]}\right\}\frac{R_1^{2n+1}\,q}{4\pi\varepsilon_1\zeta^{n+1}} \tag{C.18d}$$

with a set of effective permittivity values defined by

$$(\varepsilon_2)'_n = \varepsilon_2\left\{\frac{a^{2n+1} + (n+1)K^{(n)}}{a^{2n+1} - nK^{(n)}}\right\} \tag{C.19}$$

The coefficients $K^{(n)}$ are given by Equation (2.23) in Chapter 2, with the following substitutions: $\varepsilon_1 \to \varepsilon_2$, $\varepsilon_2 \to \varepsilon_3$. Also, $a = R_1/R_2$. Then, using Equation (2.21) for the axial derivatives of the field imposed by the point charge and $p^{(n)} = 4\pi\varepsilon_1 A_n$, the expression for the effective multipolar moments becomes

$$p^{(n)} = \frac{4\pi\varepsilon_1 R^{2n+1}}{(n-1)!}\left[\frac{(\varepsilon_2)'_n - \varepsilon_1}{n(\varepsilon_2)'_n + (n+1)\varepsilon_1}\right]\frac{\partial^{n-1}E_z}{\partial z^{n-1}} \tag{C.20}$$

The bracketed term in Equation (C.20) has the same canonic form as $K^{(n)}$, but with $(\varepsilon_2)'_n$ substituted for ε'_2. Therefore, a distinct value for the effective permittivity of the spherical dielectric shell must be used for each multipolar moment. Based on the superposition argument, Equation (C.20) is in fact generally applicable to any axisymmetric electrostatic field $E_z(z)$.

---

[1] The coefficients B_n, C_n, and D_n are useful in calculating the field within the layer and the core of the particle.

Appendix D

Transient response of ohmic dielectric sphere to a suddenly applied DC electric field

The transient response of an ohmic dielectric sphere to a suddenly applied electric field provides physical insight concerning the Maxwell–Wagner charge relaxation mechanism. Consider once again the problem illustrated in Figure 2.2 of a homogeneous sphere with permittivity ε_2, conductivity σ_2, and radius R in a fluid with permittivity ε_1 and ohmic conductivity σ_1. Assume that, at $t = 0_+$, a uniform electric field is suddenly turned on.

$$\bar{E}(t) = E_0 \hat{z} u(t) \tag{D.1}$$

where $u(t)$ is the unit step function. The solutions for the electroquasistatic potential inside and outside the sphere take the form of Equations (2.8a,b); however, now the coefficients are functions of time, that is, $A = A(t)$ and $B = B(t)$. The boundary conditions are Equations (2.9a), (2.25), and (2.26). Combining these equations yields a set of coupled differential equations.

$$R^{-3}A(t) + B(t) = E_0 u(t) \tag{D.2a}$$

$$2R^{-3}A(t)\cos\theta - \frac{\sigma_2}{\sigma_1}B(t)\cos\theta + \frac{1}{\sigma_1}\frac{\partial\sigma_f}{\partial t} = -E_0 u(t)\cos\theta \tag{D.2b}$$

$$2R^{-3}A(t)\cos\theta - \frac{\varepsilon_2}{\varepsilon_1}B(t)\cos\theta + \frac{\sigma_f}{\sigma_1} = -E_0 u(t)\cos\theta \tag{D.2c}$$

Because we need only the effective moment, it is sufficient to solve for the time-dependent coefficient $A(t)$. The solution for $p_{\text{eff}}(t \geq 0) = 4\pi\varepsilon_1 A(t)$ reveals the nature of Maxwell–Wagner surface polarization (Jones, 1979b).

$$p_{\text{eff}}(t) = 4\pi\varepsilon_1 R^3\left(\frac{\sigma_2 - \sigma_1}{\sigma_2 + 2\sigma_1}\right)E_0[1 - \exp(-t/\tau_{\text{MW}})]$$

$$+ 4\pi\varepsilon_1 R^3\left(\frac{\varepsilon_2 - \varepsilon_1}{\varepsilon_2 + 2\varepsilon_1}\right)E_0\exp(-t/\tau_{\text{MW}}) \tag{D.3}$$

where

$$\tau_{MW} = \frac{\varepsilon_2 + 2\varepsilon_1}{\sigma_2 + 2\sigma_1} \tag{D.4}$$

is the relaxation time constant associated with the accumulation of free charge $\sigma_f(t)$ at the surface of the sphere.

It is instructive to note the limiting expressions for the effective moment at times very short and very long compared to this time constant.

$$p_{eff}(t) = \begin{cases} 4\pi\varepsilon_1 R^3 \left(\dfrac{\varepsilon_2 - \varepsilon_1}{\varepsilon_2 + 2\varepsilon_1}\right) E_0, \ t \ll \tau_{MW} \\[4mm] 4\pi\varepsilon_1 R^3 \left(\dfrac{\sigma_2 - \sigma_1}{\sigma_2 + 2\sigma_1}\right) E_0, \ t \gg \tau_{MW} \end{cases} \tag{D.5}$$

These limits conform to physical intuition. On the short time scale ($t \ll \tau_{MW}$), no free charge has accumulated and the effective moment is that of an insulating dielectric sphere as examined in Section 2.2B. On the long time scale ($t \gg \tau_{MW}$), surface charge has accumulated and p_{eff} is governed by DC conduction, as discussed in Section 2.2C.

Note that the time-dependent effective dipole moment can actually change sign as a function of time. This feature may be exploited in schemes for the continuous separation of minerals based on differences of dielectric constant and conductivity (Benguigui and Lin, 1984).

In the special case where the charge relaxation times of the suspension medium and the particle are equal, that is, $\varepsilon_1/\sigma_1 = \varepsilon_2/\sigma_2 \ (\equiv \tau)$ no accumulation of unpaired (free) charge can occur at the particle's surface. Without this surface charge, there is no Maxwell–Wagner interfacial polarization and the effective dipole moment is constant in time. From Equation (D.3) for $\tau_1 = \tau_2$ only, we obtain

$$p_{eff}(t) = 4\pi\varepsilon_1 R^3 \left(\frac{\varepsilon_2 - \varepsilon_1}{\varepsilon_2 + 2\varepsilon_1}\right) E_0 = 4\pi\varepsilon_1 R^3 \left(\frac{\sigma_2 - \sigma_1}{\sigma_2 + \sigma_1}\right) E_0 \tag{D.6}$$

Any preexisting volume charge distribution will then relax according to the conventional charge relaxation law, $\exp(-t/\tau)$, without temporary charge accumulation at the surface (Woodson and Melcher, 1968, Sec. 7.2).

Appendix E

Relationship of DEP and ROT spectra

A. Time-average force and torque expressions

The time-average dielectrophoretic force on a lossy dielectric sphere in a linearly polarized sinusoidal time-varying electric field of rms magnitude E_{rms} is

$$\langle \overline{F}_{DEP}(t) \rangle = 2\pi\varepsilon_1 R^3 \mathrm{Re}[\underline{K}(\omega)] \nabla E_{rms}^2 \tag{2.46}$$

where $\underline{K}(\omega)$ is the complex Clausius–Mossotti function defined in Chapter 2 by Equation (2.30). According to this force law, regions of stronger electric field attract or repel particles depending on whether $\mathrm{Re}[\underline{K}(\omega)]$ is positive or negative, respectively. Also from Chapter 2, the torque exerted by a polarized AC electric field of magnitude E_0 upon the same dielectric sphere is

$$\langle \overline{T}^e(t) \rangle = -4\pi\varepsilon_1 R^3 \mathrm{Im}[\underline{K}(\omega)] E_0^2 \hat{z} \tag{2.50}$$

A particle will rotate with or against the rotating electric field, depending on whether the sign of $\mathrm{Im}[\underline{K}(\omega)]$ is negative or positive, respectively.

The observation that the DEP force and the rotational torque are proportional to the real and imaginary components of $\underline{K}(\omega)$, respectively, is quite important. The purpose of this appendix is to reveal the fundamental relationship that exists between $\mathrm{Re}[\underline{K}]$ and $\mathrm{Im}[\underline{K}]$ and then to take advantage of it to infer useful information about the ROT spectra from the DEP spectra and vice versa.

B. Generalized expression for $\underline{K}(\omega)$

Using the methods of Section 2.3 in Chapter 2, we can in principle model a heterogeneous sphere with any number of layers and coatings. However, as layers are added and Equations (2.34) and (C.6) are used repeatedly, the algebraic expression for $\underline{K}(\omega)$ quickly becomes too unwieldy for analysis. Nevertheless, because most interesting particles feature multiple layers, there is strong motivation to understand how $\underline{K}(\omega)$ behaves as the frequency is varied and as various system parameters are changed. To proceed in a systematic way, we perform the replacement $j\omega \to s$, where s is the complex frequency familiar from classical analog circuit analysis. This transformation is equivalent to assuming exponential time solutions of the form: $\exp(st)$. When this transformation is performed, $\underline{K} \to K(s)$ and $K(s)$ takes the form of a fractional polynomial expression with both the numerator and denominator of equal order N.

$$K(s) = K_\infty \frac{s^N + C_{N-1}s^{N-1} + C_{N-2}s^{N-2} + \ldots + C_0}{s^N + D_{N-1}s^{N-1} + D_{N-2}s^{N-2} + \ldots + D_0} \tag{E.1}$$

where $K_\infty = K(s \to \infty)$ and the coefficients C_i and D_i are real and positive. In the absence of dielectric resonances, the N real poles are limited to negative values, while the equal number of zeros can be positive or negative. In terms of the circuit analogy, $K(s)$ has the form of a transfer function, relating the imposed electric field to the excess polarization of the particle. Note that if the only loss mechanism in the particle and the suspension fluid is ohmic conduction, then N equals the number of interfaces of the particle.

In general, we cannot readily factor these polynomials to obtain algebraic expressions for the poles or zeros. Even for simple cases, the expressions obtained using computer-based algebraic manipulators are algebraically complex. Nevertheless, with the knowledge that all the poles of $K(s)$ are real and negative and that all the zeros are real, it becomes possible to rewrite Equation (E.1) in the more useful form of a partial fraction expansion.

$$K(s) = K_\infty - \frac{s_\alpha \Delta K_\alpha}{s + s_\alpha} - \frac{s_\beta \Delta K_\beta}{s + s_\beta} - \ldots - \frac{s_N \Delta K_N}{s + s_N} \tag{E.2}$$

where $s_i > 0$ for $i = \alpha, \beta, \gamma, \ldots, N$. (Note that the index values $\alpha, \beta, \gamma, \ldots$ are used here to distinguish the poles of $K(s)$ from the indices $1, 2, 3, \ldots$ needed to denote individual layers in heterogeneous particles.) Each term in Equation (E.2) represents either a Maxwell–Wagner relaxation mechanism associated with an interface or an intrinsic dielectric dispersion in one of the layers composing the particle. The low-frequency limit of $K(s)$ is

$$K_0 = K_\infty - \sum_{i=\alpha}^{N} \Delta K_i \tag{E.3}$$

Having taken advantage of circuit analysis techniques to reduce the form of $K(s)$ to continued fraction form, we now perform the reverse transformation $s \to j\omega$ to write the complex frequency-dependent $\underline{K}(\omega)$.

$$\underline{K}(\omega) = K_\infty - \frac{\Delta K_\alpha}{j\omega\tau_\alpha + 1} - \frac{\Delta K_\beta}{j\omega\tau_\beta + 1} - \cdots - \frac{\Delta K_N}{j\omega\tau_N + 1} \qquad (E.4)$$

where $\tau_i = s_i^{-1}$ for $i = \alpha, \beta, \ldots, N$. With no loss of generality, we here assume that the relaxation times are ordered as follows: $\tau_\alpha > \tau_\beta > \tau_\gamma > \ldots > \tau_N$. Algebraic expressions for the relaxation times and for the factors ΔK_α, ΔK_β, $\ldots$ are not easily obtained. Nevertheless, this form for $\underline{K}(\omega)$ is much more convenient for analytical purposes than is the Clausius–Mossotti function (Fuhr, 1986; Turcu, 1987). In fact, when the poles are widely separated, that is, $\tau_\alpha \gg \tau_\beta \gg \ldots \gg \tau_N$, we can often obtain approximate expressions for the τ_i's and ΔK_i's.

As a simple example, consider a homogeneous dielectric sphere of radius R, permittivity ε_2, and conductivity σ_2 immersed in a dielectric medium of permittivity ε_1 and conductivity σ_1. This simple system features one interface and one relaxation frequency so that we may write Equation (2.31) in the convenient form of Equation (E.4).

$$\underline{K}(\omega) = K_\infty + \frac{K_0 - K_\infty}{j\omega\tau_{MW} + 1} \qquad (E.5)$$

where

$$K_\infty \equiv \frac{\varepsilon_2 - \varepsilon_1}{\varepsilon_2 + 2\varepsilon_1} \quad \text{and} \quad K_0 \equiv \frac{\sigma_2 - \sigma_1}{\sigma_2 + 2\sigma_1} \qquad (E.6)$$

K_∞ and K_0 are the same high- and low-frequency limits of $\underline{K}(\omega)$ defined by Equations (2.33a,b) in Chapter 2. Equation (E.5) matches the classical form of the Debye relaxation equation for the complex dielectric constant (Coelho, 1979).

C. Argand diagrams of $\underline{K}(\omega)$

One excellent way to examine the frequency response of a particle is to map the real and imaginary parts of $\underline{K}(\omega)$ onto the complex plane as the frequency is varied. Such maps, called Argand plots, have been used as a tool for investigating the rotational spectra of biological cells and solid particles (Fuhr, 1986).

C.1 Theory for homogeneous sphere

The essential properties of Argand plots of $\underline{K}(\omega)$ are revealed readily using a homogeneous sphere as the model particle. We proceed by recasting Equation (E.4) in the following useful form (Jones and Kaler, 1990).

$$\underline{K}(\omega) = \frac{1}{2}(K_0 + K_\infty) + \frac{1}{2}(K_0 - K_\infty)\exp\left(-j2\tan^{-1}\omega\tau_{MW}\right) \qquad (E.7)$$

where $\tau_{MW} = (\varepsilon_2 + 2\varepsilon_1)/(\sigma_2 + 2\sigma_1)$ is the Maxwell–Wagner polarization time constant already defined by Equation (2.32) in Section 2.3A. From Equation (E.7), it is evident that the locus of $\underline{K}(\omega)$ for increasing ω is a semicircle prescribed in the clockwise direction that starts on the real axis at K_0 and ends at K_∞, also on the real axis. It will be obvious that the center of this circle is on the real axis at $(K_0 + K_\infty)/2$ and that the radius is $|K_0 - K_\infty|/2$.

Figure E.1 shows typical loci of $\underline{K}(\omega)$ obtained from Equation (E.7) for a homogeneous sphere. These plots exemplify a set of general rules for all Argand diagrams of $\underline{K}(\omega)$: (i) the locus consists of one (or more) semicircles entirely contained in the upper or lower half-plane; (ii) each semicircle is traced in a clockwise direction as the frequency ω is increased; (iii) all loci are confined within a circle of radius 0.75 centered on the real axis at 0.25. For convenience, each of the four quadrants is identified with respect to ±DEP and ±ROT.

There is a direct correspondence between the ROT and DEP spectra of a particle and its $\underline{K}(\omega)$ locus. The highest (or lowest) point of the semicircular arc represents a peak in the ROT spectrum. Whenever the real part of $\underline{K}$ increases with frequency, there is a negative rotational peak, whereas whenever the real part of $\underline{K}$ decreases, a positive rotational peak occurs. Note how the rotational peaks always occur at the midpoint of the change in $\text{Re}[\underline{K}(\omega)]$. These rules for locating the locus of $\underline{K}(\omega)$ are illustrated in Figures E.2a and b, which plot $\text{Re}(\underline{K})$ and $\text{Im}(\underline{K})$ versus frequency for the same parameters used in Figure E.1. Transitions in $\text{Re}[\underline{K}(\omega)]$ denote changes in the DEP force or, in the case of DEP levitation, an adjustment in the voltage required to maintain a particle at fixed position (Kaler and Jones, 1990). A change in the sign of $\text{Re}[\underline{K}(\omega)]$ indicates that the requirement for levitation changes from passive to feedback-controlled or vice versa.

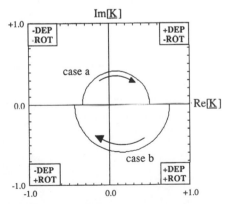

Fig. E.1 Argand plots of $\underline{K}(\omega)$ for homogeneous sphere: case (a) $K_\infty > K_0$, $\varepsilon_2/\varepsilon_1 = 4.0$, $\sigma_2/\sigma_1 = 0.25$; case (b) $K_\infty < K_0$, $\varepsilon_2/\varepsilon_1 = 0.1$, $\sigma_2/\sigma_1 = 10.0$. The four quadrants are identified as to ±DEP and ±ROT.

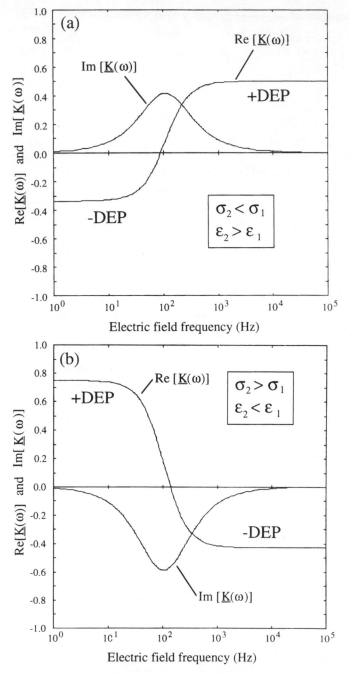

Fig. E.2 Re[$\underline{K}$] & Im[$\underline{K}$] spectra using the parameters of Figure E.1.
(a) $\varepsilon_1/\varepsilon_0 = 2.5$, $\varepsilon_2/\varepsilon_0 = 10.0$, $\sigma_1 = 4 \cdot 10^{-8}$ S/m, $\sigma_2 = 10^{-8}$ S/m, $R = 5 \, \mu$m.
(b) $\varepsilon_1/\varepsilon_0 = 10.0$, $\varepsilon_2/\varepsilon_0 = 1.0$, $\sigma_1 = 10^{-8}$ S/m, $\sigma_2 = 10^{-7}$ S/m, $R = 5 \, \mu$m.

C.2 Generalization for widely separated relaxation frequencies

The above rules for Argand plots of $\underline{K}(\omega)$ are quite general and apply readily to any multilayered particle, as long as the various relaxation frequencies of the particle are widely separated. In this case, the trajectory of $\underline{K}(\omega)$ will consist of a set of linked semicircles – one for each relaxation process – traced in the clockwise direction as ω increases. To demonstrate this generality, consider Equation (E.4) for $\underline{K}(\omega)$ under the restriction of widely separated relaxation frequencies, that is, $\tau_\alpha \gg \tau_\beta \gg \ldots \gg \tau_N$. For a range of frequencies centered on the nth relaxation, such that $(\tau_{n-1})^{-1} \gg \omega \gg (\tau_{n+1})^{-1}$, we may write

$$\underline{K}(\omega) \approx K_\infty - \sum_{i=n+1}^{N} \Delta K_i - \frac{\Delta K_n}{j\omega\tau_n + 1} \tag{E.8}$$

Comparison of Equation (E.8) to Equation (E.5) reveals their similar form. Thus

$$\underline{K}(\omega) \approx \frac{1}{2}(K_{n-1} + K_n) + \frac{1}{2}(K_{n-1} - K_n)\exp(-j2\tan^{-1}\omega\tau_n) \tag{E.9}$$

where K_{n-1} and K_n represent the asymptotic real values of $\underline{K}(\omega)$ on either side of the relaxation occurring at $\omega = 1/\tau_n$. Because Equation (E.9) is identical to Equation (E.7), the trajectory of $\underline{K}(\omega)$ near each relaxation frequency takes the form of a semicircular arc starting and ending on the real axis. All the rules listed for the simpler case of the homogeneous sphere apply here as well. When the relaxation frequencies are close together, the semicircles become superimposed and the trajectories are distorted, making individual peaks harder to discern.

Care must be exercised in use of Argand diagrams for particles with multiple relaxation processes because physical realizability imposes certain restrictions on the allowable trajectories for $\underline{K}(\omega)$ (Kaler et al., 1992; Wang et al., 1992). Apparent violations of these restrictions by experimentally obtained DEP or electrorotation data imply serious deficiencies of the model chosen to represent the particle's behavior in an electric field.

C.3 Kramers–Krönig relations

As a consequence of the linear relationship between $\bar{p}_{\text{eff}}$ and $\bar{E}$, i.e., Equation (2.29), $\text{Re}[\underline{K}(\omega)]$ and $\text{Im}[\underline{K}(\omega)]$ must be analytically interrelated through the well-known Kramers–Krönig relations (Pastushenko et al.,1985; Landau and Lifshitz, 1960, Sec. 62). These relationships are in the form of a pair of integrals.

$$\text{Re}[\underline{K}(\omega)] - K_\infty = \frac{2}{\pi} \int_0^\infty \frac{x\,\text{Im}[\underline{K}(x)]}{x^2 - \omega^2} dx \tag{E.10}$$

$$\text{Im}[\underline{K}(\omega)] - K_\infty = -\frac{2\omega}{\pi} \int_0^\infty \frac{\text{Re}[\underline{K}(x)] - K_\infty}{x^2 - \omega^2} dx \tag{E.11}$$

where $K_\infty = \text{Re}[K(\omega \to \infty)]$. Equations (E.10) and (E.11) explain why rotational peaks are always accompanied by a change in the dielectrophoretic force and vice versa. The Kramers–Krönig equations provide a unifying theoretical framework for the dielectrophoretic and electrorotational spectra of particles exhibiting any number of first-order dielectric dispersion mechanisms.

With the intimate relationship of $\text{Re}[\underline{K}]$ and $\text{Im}[\underline{K}]$ now demonstrated, one might argue that the DEP levitation method of Section 3.4 and the electrorotation method of Section 4.3 provide identical information concerning a particle such as a biological cell. This statement is true in principle but quite misleading in any practical sense. Because of the widely differing sensitivities to cell modeling parameters of the real and imaginary components of $\underline{K}$, the electrorotation and DEP levitation techniques are often suited best to different sorts of measurements (Foster et al.,1992). It is best to say that the two techniques fill complementary roles in the investigation of individual cell and particle properties.

D. Illustrative examples

The Argand diagrams provided below exemplify typical behavior for heterogeneous (layered) spherical particles having more than one Maxwell–Wagner relaxation frequency. For reference, Cole–Cole plots of complex effective particle permittivity $\underline{\varepsilon}_2'$ are also plotted in some cases. In addition, examples of some simple models for biological cells (yeast cells and protoplasts) are included.[1]

D.1 Multilayered spheres

Argand diagrams achieve significant interpretive value in representing the dispersion of multilayered spherical shells consisting of ohmic dielectrics. If the relaxation frequencies of the shell are widely separated, the $\underline{K}(\omega)$ locus appears as a set of linked semicircles with each circle representing a Maxwell–Wagner relaxation mechanism. Figures E.3a and b show Argand plots for two different three-layered spherical particles. The parameters of the particles are provided in the figure caption. To obtain these plots, we have replaced $\underline{\varepsilon}_2$ in $\underline{K}(\omega)$ by the effective permittivity $\underline{\varepsilon}_2'$ of the multilayered particle structure. Indicative that the Maxwell–Wagner relaxation frequencies are widely separated, each Argand locus consists of three distinct semicircles. Note how the diagrams are influenced by the relative values of the intrinsic charge relaxation time constants for the suspension medium (τ_1) and for the three layers of the particle (τ_2, τ_3, τ_4), where $\tau_i = \varepsilon_i/\sigma_i$.

Argand plots of $\underline{K}(\omega)$ are distinct from Cole–Cole plots in several important respects. First, the $\underline{K}(\omega)$ and $\varepsilon(\omega) = \varepsilon'(\omega) - j\varepsilon''(\omega)$ trajectories rotate in opposite

---

[1] All Argand and Cole–Cole plots appearing in this book were produced using an interactive Macintosh Pascal® computer program originated by Saulei Chan at the University of Rochester in 1988.

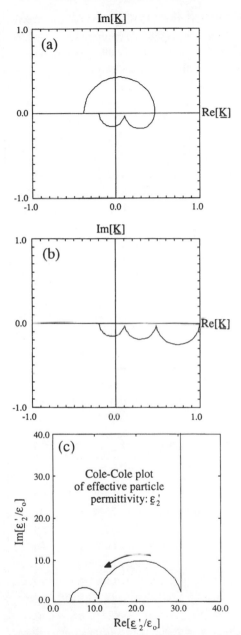

Fig. E.3 Argand and Cole-Cole diagrams for the layered spherical shell shown in Figure 2.6, illustrating the influence of relative intrinsic charge relaxation times upon spectra.

(a) Argand plot for $\tau_4 \ll \tau_3 \ll \tau_1 \ll \tau_2$: $\varepsilon_1/\varepsilon_0 = 8.0$, $\sigma_1 = 3 \cdot 10^{-6}$ S/m; $R_1 = 12$ μm; $\varepsilon_2/\varepsilon_0 = 6.0$, $\sigma_2 = 10^{-7}$ S/m; $R_2 = 10$ μm; $\varepsilon_3/\varepsilon_0 = 4.0$, $\sigma_3 = 10^{-4}$ S/m; $R_3 = 8$ μm; $\varepsilon_4/\varepsilon_0 = 2.0$, $\sigma_4 = 10^{-1}$ S/m.

(b) Argand plot for $\tau_4 \ll \tau_3 \ll \tau_2 \ll \tau_1$: all particle parameters are identical to (a), except that $\sigma_1 = 10^{-9}$ S/m.

(c) Cole-Cole plot for the spherical shell with parameters defined in (a).

directions. This difference is a consequence of the convention defining $\varepsilon''(\omega)$ to be the negative imaginary component of $\underline{\varepsilon}(\omega)$. Second, $\underline{K}(\omega)$ trajectories can be above or below the real axis, while Cole–Cole plots are always on or above the real axis. This is because the lossy nature of all dielectrics demands that $\varepsilon''(\omega)$ > 0. No such restriction applies to $\underline{K}(\omega)$, which should be thought of as a transfer function rather than a self-admittance. Compare the Argand plot of $\underline{K}(\omega)$ to the Cole–Cole plot of $\varepsilon_2'(\omega)$ in Figures E.3a and E.3c.

D.2 Biological cells

Several examples of simple models for biological cells are provided here to show how changes in the modeling parameters can influence the DEP and ROT spectra. Consider the simple spherical model of a protoplast (or mammalian erythrocyte), shown in Figure 3.4a of Chapter 3. The cell, of radius R, consists of homogeneous cytoplasm of permittivity ε_2 and ohmic conductivity σ_2 enclosed by a membrane of capacitance per unit area c_m and is suspended in a medium of permittivity ε_1 and ohmic conductivity σ_1. The effect upon the DEP and ROT spectra of the medium conductivity is illustrated in Figure E.4a. Note that these Argand diagrams disguise the influence of parameters such as σ_1 and c_m upon peak rotation frequencies. This information can be obtained readily but is not provided in the diagrams due to lack of space.

Figure E.4b shows the influence of finite transmembrane conductance per unit area g_m (S/cm^2) upon the ROT and DEP spectra. This conductance provides a simple way to represent the effect of a distribution of transmembrane pores

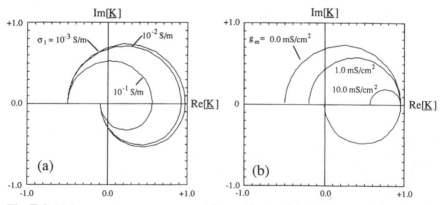

Fig. E.4 (a) Argand diagrams for the model protoplast of Figure 3.4a for several values of suspension conductivity: cytoplasm: $\varepsilon_c/\varepsilon_0 = 60.0$, $\sigma_c = 0.5$ S/m, $R = 2.0$ μm; membrane: $c_m = 1.0$ μF/cm^2; suspension medium: $\varepsilon_1/\varepsilon_0 = 78.0$, $\sigma_1 = 10^{-3}$, 10^{-2}, 10^{-1} S/m. (b) Argand diagrams of a model protoplast with varied membrane transconductance $g_m = 0$, 1.0, and 10.0 mS/cm^2 and $\sigma_1 = 10^{-4}$ S/m. All other parameters are identical to those in (a).

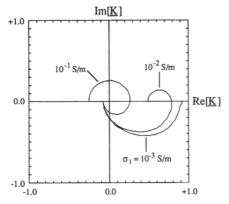

Fig. E.5 Argand diagrams for the model walled cell of Figure 3.3a for several values of suspension conductivity: cytoplasm: $\varepsilon_c/\varepsilon_0 = 60.0$, $\sigma_c = 0.5$ S/m, $R = 2.0$ µm; membrane: $c_m = 1.0$ µF/cm^2, $g_m = 0$; cell wall: $\varepsilon_w/\varepsilon_0 = 65.0$, $\sigma_w = 0.1$ S/m, $R_w = 2.5$ µm; suspension medium: $\varepsilon_1/\varepsilon_0 = 78.0$, $\sigma_1 = 10^{-3}$, 10^{-2}, 10^{-1} S/m.

(which permit the flow of DC current through the membrane) on the low-frequency behavior.

Figure E.5 plots $\underline{K}(\omega)$ loci for a model yeast cell, consisting of a cytoplasm, lossless membrane, and outer cell wall, for several different values of suspension conductivity σ_1. Cell parameters typical of the common yeast *Sacch. cerevisiae* have been used here. Note that the conductive cell wall effectively shields the membrane and cytoplasm from the probing AC field at lower frequencies. At the lowest value of the medium conductivity σ_1, only one rotational peak is evident because the two system poles are closely spaced.

E. Conclusion

Argand diagrams of the complex polarization coefficient $\underline{K}(\omega)$ provide compact representations for frequency-dependent electromechanics of particles and biological cells in suspension. Though they disguise the absolute values of relaxation frequency breakpoints, such plots facilitate interpretation of the influences of important parameters upon the rotational and dielectrophoretic levitation spectra of cells. Gimsa et al. (1991) and Huang et al. (1992) have successfully plotted Re[$\underline{K}$] and Im[$\underline{K}$] data obtained from separate DEP and ROT experiments in this format. To date, however, no one has devised the means for direct, simultaneous measurement of both the real and imaginary components of $\underline{K}(\omega)$ for single particles. Even so, Argand diagrams have clear utility in the investigation and interpretation of complex models for biological cells and many other types of small particles in aqueous suspension.

Appendix F

General multipolar theory

The linear multipoles first introduced in Section 2.2E and then employed to formulate higher-order force corrections in Section 2.4B are applicable only when the electric field is axisymmetric. The problem of a dielectric particle in a completely arbitrary, nonuniform field requires a more general version of the effective multipolar theory. Consider the general set of moments represented by Figure F.1. Any moment of order n is produced by starting with a pair of opposite-sign multipoles of order $n-1$ that are displaced from one another by the vector

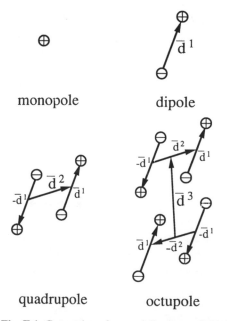

Fig. F.1 Generation of general dipole, quadrupole, and octupole starting from monopolar charges and using the vector displacements $\bar{d}_1, \bar{d}_2, \bar{d}_3$, respectively. Compare these to the linear multipoles depicted in Figure B.2.

distance $\bar{d}_n$. By convention, this vector always points from the negative to the positive moment. Using the same arguments employed to arrive at Equation (2.4) for the force exerted upon a dipole, one may show that the force on the nth general multipole $\overset{\underset{\cdot\cdot\cdot}{=}}{p}{}^n$ is

$$\bar{F}_n = \frac{1}{n!}\overset{\underset{\cdot\cdot\cdot}{=}}{p}{}^n[\bullet]^n(\nabla)^n\bar{E} \tag{F.1}$$

We employ dyadic notation here to represent the tensor nature of the higher-order multipoles (Bird et al., 1977). Thus, $(\nabla)^n\Phi = \nabla\nabla\cdots\nabla\Phi$ is a symmetric tensor of rank n, while $[\bullet]^n$ represents n dot product operations. The dyadic tensor moment relates to the original dipole vector definition $\bar{p} \equiv q\bar{d}$ through

$$\overset{\underset{\cdot\cdot\cdot}{=}}{p}{}^n = n!q\,(\bar{d}_n\bar{d}_{n-1}\cdots\bar{d}_2\bar{d}_1) \tag{F.2}$$

The induced potentials are linearly related to these moments.

$$\Phi_n = \frac{(-1)^n}{4\pi n!\varepsilon_1}\overset{\underset{\cdot\cdot\cdot}{=}}{p}{}^n[\bullet]^n(\nabla)^n\!\left(\frac{1}{r}\right) \tag{F.3}$$

In the general multipolar theory, Equation (F.3) takes the place of Equation (2.7) and is used to identify the tensor moment $\overset{\underset{\cdot\cdot\cdot}{=}}{p}{}^n$.

For a spherical dielectric particle of radius R and permittivity ε_2 in a fluid of permittivity ε_1, the tensor moments are related to gradients of the imposed electrostatic field vector (Washizu and Jones, 1994).

$$\overset{\underset{\cdot\cdot\cdot}{=}}{p}{}^n = \frac{4\pi\varepsilon_1 R^{2n+1}n}{(2n-1)!!}K^{(n)}(\nabla)^{n-1}\bar{E} \tag{F.4}$$

where $(2n-1)!! \equiv (2n-1)\bullet(2n-3)\bullet\ldots\bullet5\bullet3\bullet1$ and $K^{(n)}$ is defined by Equation (2.23). Equation (F.4) is the multipolar generalization of Equation (2.22). When this equation is combined with (F.1), we obtain an expression that can be used to calculate the ponderomotive force on a spherical dielectric particle to any degree of desired accuracy in any electrostatic field $\bar{E}$. The first three terms, accounting for dipolar, quadrupolar, and octupolar contributions, are (Washizu and Jones, 1994).

$$\bar{F}_{\text{total}} = 2\pi\varepsilon_1\nabla\left\{K^{(1)}R^3[\bar{E}\bullet\bar{E}] + \frac{K^{(2)}R^5}{3}[\nabla\bar{E}:\nabla\bar{E}] + \right.$$

$$\left.\frac{K^{(3)}R^7}{30}[\nabla\nabla\bar{E}\vdots\nabla\nabla\bar{E}] + \cdots\right\} \tag{F.5}$$

Equation (F.5) may be thought of as the general multipolar equivalent of Equation (2.43), which accounts only for linear multipoles in an axisymmetric field.

In most applications of DEP, the conventional dipole approximation is adequate for the purposes of estimating the force. On the other hand, recent passive DEP levitation experiments performed with new classes of salient electrode structures (Washizu and Nakada, 1991; Fuhr et al., 1992) demonstrate that the quadrupolar moment is responsible for the levitation effect (Washizu et al., 1993a).[1]

---

[1] Some of these new structures feature a zero of the electric field along the central axis. Because the net dipole moment of a particle located on this axis must be zero, the dipole force cannot be responsible for the levitation effect.

Appendix G

Induced effective moment of dielectric ellipsoid

We can identify the induced effective moment of a dielectric ellipsoid readily by examining the limit of the electrostatic potential at a point far from the ellipsoid, where ellipsoidal coordinates degenerate into spherical coordinates. Consider an ellipsoid with semi-axes a, b, and c and dielectric permittivity ε_2 suspended in a fluid of permittivity ε_1. The axis a is aligned parallel to the uniform imposed electrostatic field component E_x. We may express the solution for the electrostatic potential external to the ellipsoid in integral form (Stratton, 1941, Sec. 3.27).

$$\Phi_1 = \Phi_0 + \frac{Z_2}{Z_1}\Phi_0 \int_\zeta^\infty \frac{ds}{(s+a^2)R_s} \tag{G.1}$$

where $\Phi_0 = -E_x x = -E_x r \cos\theta$ is the imposed electrostatic potential, ξ is an elliptical coordinate, and $R_s = \sqrt{(s+a^2)(s+b^2)(s+c^2)}$. Application of the boundary conditions provides a relationship between the coefficients Z_1 and Z_2.

$$Z_2 = \frac{-\dfrac{abc}{2}(\varepsilon_2 - \varepsilon_1)Z_1}{\varepsilon_1 + (\varepsilon_2 - \varepsilon_1)\dfrac{abc}{2}\displaystyle\int_0^\infty \frac{ds}{(s+a^2)R_s}} \tag{G.2}$$

Using Equation (G.2) in Equation (G.1) with the depolarization factor L_x defined by Equation (5.4), and then taking the limit $\xi \gg a^2$, b^2, and c^2 to reduce the elliptic integral, we obtain a simplified expression for the second term in Equation (G.1), the induced component of the electrostatic potential.

$$\Phi_1^+ \approx -\frac{\dfrac{abc}{2}(\varepsilon_2 - \varepsilon_1)(-E_x r \cos\theta)\displaystyle\int_\xi^\infty \frac{ds}{s^{5/2}}}{\varepsilon_1 + (\varepsilon_2 - \varepsilon_1)L_x} \tag{G.3}$$

Far from the ellipsoid, $\xi \approx r^2$ and the induced potential reduces to the familiar form of the dipole solution in spherical coordinates.

$$\Phi_1^+ \approx -\frac{\frac{abc}{3}(\varepsilon_2 - \varepsilon_1)E_x}{\varepsilon_1 + (\varepsilon_2 - \varepsilon_1)L_x}\frac{\cos\theta}{r^2} \tag{G.4}$$

Using Equations (G.4) and (2.7), we may now identify the x component of the effective dipole moment due to the dielectric ellipsoid.

$$(p_{\mathrm{eff}})_x \approx \frac{4\pi abc}{3}\varepsilon_1\left[\frac{\varepsilon_2 - \varepsilon_1}{\varepsilon_1 + (\varepsilon_2 - \varepsilon_1)\,L_x}\right]E_x \tag{G.5}$$

The other components of the effective moment, $(p_{\mathrm{eff}})_y$ and $(p_{\mathrm{eff}})_z$, are of similar form.

References

Allan, R. S., and S. G. Mason, "Particle behavior in shear and electric fields: I. Deformation and burst of fluid drops," *Proc. Royal Soc. (London)*, vol. A267, 1962, 45–61.

Alward, J., and W. Imaino, "Magnetic forces on monocomponent toner," *IEEE Trans. Mag.*, vol. MAG–22, 1986, 128–134.

Andres, U., *Magnetohydrodynamic and magnetohydrostatic methods of mineral separation.* New York: John Wiley and Sons, 1976.

Andres, U., and W. O'Reilly, "Separation of minerals by selective magnetic filtration," *Powder Technol.*, vol. 69, 1992, 279–284.

Arno, R., Atti Reale Accad. dei Lincei, *Rendiconti*, vol. 1, 1892, 284.

Arnold, W. M., "Analysis of optimum electro-rotation technique," *Ferroelectrics*, vol. 86, 1988, 225–244.

Arnold, W. M., R. K. Schmutzler, S. Al-Hasani, D. Krebs, and U. Zimmermann, "Differences in membrane properties between unfertilized and fertilized single rabbit oocytes demonstrated by electrorotation. Comparison with cells from early embryos," *Biochim. Biophys. Acta*, vol. 979, 1989, 142–146.

Arnold, W. M., H. P. Schwan, and U. Zimmermann, "Surface conductance and other properties of latex particles measured by electrorotation," *J. Phys. Chem.*, vol. 91, 1987, 5093–5098.

Arnold, W. M., B. Wendt, U. Zimmermann, and R. Korenstein, "Rotation of a single swollen thylakoid vesicle in a rotation electric field. Electrical properties of the photosynthetic membrane and their modification by ionophores, lipophilic ions, and pH," *Biochim. Biophys. Acta*, vol. 813, 1985, 117–131.

Arnold, W. M., and U. Zimmermann, "Rotating-field-induced rotation and measurement of the membrane capacitance of single mesophyll cells of *Avena sativa*," *Z. Naturforsch.*, vol. 37, 1982, 908–915.

Arp, P. A., R. T. Foister, and S. G. Mason, "Some electrohydrodynamic effects in fluid dispersions," *Advances in Colloid Interface Sci.*, vol. 12, 1980, 295–356.

Arp. P. A., and S. G. Mason, "Chains of spheres in shear and electric fields," *Colloid Polymer Sci.*, vol. 255, 1977, 1165–1173.

Asami, K., T. Hanai, and N. Koizumi, "Dielectric approach to suspensions of ellipsoidal particles covered with a shell, in particular reference to biological cells," *Japan. J. Appl. Phys.*, vol. 19, 1980, 359–365.

Bahaj, A. S., and A. G. Bailey, "Dielectrophoresis of small particles," Proc. of IEEE/IAS Annual Meeting, Cleveland, Ohio, October 1979, 154–157.

Batchelder, J. S., "Dielectrophoretic manipulator," *Rev. Sci. Instrum.*, vol. 54, 1983, 300–302.

Bates, L. F., *Modern magnetism*, 4th ed. Cambridge: Cambridge University Press, 1961.

Becker, R., *Electromagnetic fields and interactions.* Blaisdell: London, 1964.

Bedeaux, D., M. M. Wind, and M. A. van Dijk, "The effective dielectric constant of a dispersion of clustering spheres," *Z. Phys. B – Cond. Matter,* vol. 68, 1987, 343–354.

Benguigui, L., and I. J. Lin, "More about the dielectrophoretic force," *J. Appl. Phys.,* vol. 53, 1982, 1141–1141.

Benguigui, L., and I. J. Lin, "Phenomenological aspects of particle trapping by dielectrophoresis," *J. Appl. Phys.,* vol. 56, 1984, 3294–3297.

Benguigui, L., and I. J. Lin, "Dielectrophoresis in two dimensions," *J. Electrostatics,* vol. 21, 1988, 205–213.

Benguigui, L., A. L. Shalom, and I. J. Lin, "Influence of the sinusoidal field frequency on dielectrophoretic capture of a particle on a rod," *J. Phys. D: Appl. Phys.,* vol. 19, 1986, 1853–1861.

Bird, R., B., R. C. Armstrong, and O. Hassager, *Dynamics of polymeric liquids: Fluid mechanics* (vol. 1). New York: Wiley, 1977.

Block, H., J. P. Kelly, A. Qin, and T. Watson, "Materials and mechanisms in electrorheology," *Langmuir,* vol. 6, 1990, 6–14.

Born, M. "Über die Beweglichkeit der elektrolytischen Ionen," *Z. Phys.,* vol. 1, 1920, 221–249.

Bozzo, R., G. Coletti, P. Molfino, and G. Molinari, "Electrode systems for dielectric strength tests controlling the electro-dielectrophoretic effect," *IEEE Trans. Elect. Insul.,* vol. EI-20, 1985a, 343–348.

Bozzo, R., P. Girdinio, F. Lazzeri, and A. Viviani, "Parameter identification for a model of impurity particle motion in a dielectric fluid," *IEEE Trans. Elect. Insul.,* vol. EI-20, 1985b, 389–393.

Brandt, E. H., "Rigid levitation and suspension of high-temperature superconductors by magnets," *Amer. J. Phys.,* vol. 58, 1990, 43–49.

Burt, P. H., T. A. K. Al-Ameen, and R. Pethig, "An optical dielectrophoresis spectrometer for low-frequency measurements on colloidal suspensions," *J. Phys. E: Instrum.,* vol. 22, 1989, 952–957.

Chen, Y., A. F. Sprecher, and H. Conrad, "Electrostatic particle-particle interactions in electrorheological fluids," *J. Appl. Phys.,* vol. 70, 1991, 6796–6803.

Coelho, R., *Physics of Dielectrics.* Amsterdam: Elsevier Scientific Publishing Co., 1979.

Colver, G. M., "Dynamics of an electric (particulate) suspension," *Adv. Mech. Flow Granular Mat.* (M. Shahinpoor, ed.), vol. 1, 1983, 355–371.

Davis, L. C., "Finite-element analysis of particle-particle forces in electrorheological fluids," *Appl. Phys. Lett.,* vol. 60, 1992, 319–321.

Davis, M. H., "Two charged spherical conductors in a uniform electric field: Forces and field strength," *Quart. J. Mech. Appl. Math.,* vol. XVII, 1964, 499–511.

Davis, M. H., "Electrostatic field and force on a dielectric sphere near a conducting plane," *Amer. J. Phys.,* vol. 37, 1969, 26–29.

Dietz, P. W., "Electrofluidized bed mechanics," Ph. D. dissertation, Mass. Inst. of Technol., Cambridge, Mass., 1976.

Dietz, P. W., and J. R. Melcher, "Field controlled charge and heat transfer involving macroscopic charged particles in liquids," *ASME J. Heat Transfer,* vol. 97, 1975, 429–434.

Dietz, P. W., and J. R. Melcher, "Interparticle electrical forces in packed and fluidized beds," *Ind. Eng. Chem. Fundam.,* vol. 17, 1978, 28–32.

Dukhin, S. S., "Dielectric properties of disperse systems," in *Surface Colloid Sci.* (E. Matijevic, ed.), vol. 3, 1971, 83–164.

Engelhardt, H., H. Gaub, and E. Sackmann, "Viscoelastic properties of erythrocyte membranes in high-frequency electric fields," *Nature,* vol. 307, 1984, 378–380.

Feeley, C. M., and F. McGovern, "Dielectrophoresis of bubbles in isomotive electric fields," *J. Phys. D.: Appl. Phys.*, vol. 21, 1988, 1251–1254.

Foster, K. R., F. A. Sauer, and H. P. Schwan, "Electrorotation and levitation of cells and colloidal particles," *Biophys. J.*, vol. 63, 1992, 180–190.

Fowlkes, W. Y., and K. S. Robinson, "The electrostatic force on a dielectric sphere resting on a conducting substrate," in *Particles on surfaces: Detection, adhesion, and removal* (K. L. Mittal, ed.). New York: Plenum Press, 1988, 143–155.

Freitag, R., Schügerl, W. M. Arnold, and U. Zimmermann, "The effect of osmotic and mechanical stresses and enzymatic digestion on the electro-rotation of insect cells (*Spodoptera frugiperda*)," *J. Biotech.* vol. 11, 1989, 325–336.

Friend, A. W., E. D. Finch, and H. P. Schwan, "Low frequency electric field induced changes in the shape and motility of Amoebas," *Science*, vol. 187, 1975, 357–359.

Fröhlich, H., *Theory of Dielectrics*, Oxford: University Press, 1958.

Fuhr, G. R., "Über die Rotation dielektrischer Körper in rotierenden Feldern," Habil. Dissertation, Humboldt–Universität zu Berlin, 1986.

Fuhr, G., W. M. Arnold, R. Hagedorn, T. Müller, W. Benecke, B. Wagner, and U. Zimmermann, "Levitation, holding, and rotation of cells within traps made by high-frequency fields," *Biochim. Biophys. Acta*, vol. 1108, 1992, 215–223.

Fuhr, G., R., and R. Hagedorn, "Dielectric rotation and oscillation – a principle in biological systems?" *Studia Biophys.*, vol. 121, 1987, 25–36.

Fuhr, G., R. Hagedorn, and H. Göring, "Separation of different cell types by rotating electric fields," *Plant Cell Physiol.*, vol. 26, 1985, 1527–1531.

Fuhr, G., T. Müller, and R. Hagedorn, "Reversible and irreversible rotating field-induced membrane modifications," *Biochim. Biophys. Acta*, vol. 980, 1989, 1–8.

Füredi, A. A., I. Ohad, "Effects of high-frequency electric fields on the living cell," *Biochim. Biophys. Acta*, vol. 79, 1964, 1–8.

Füredi, A. A., and R. C. Valentine, "Factors involved in the orientation of microscopic particles in suspensions influenced by radio frequency fields," *Biochim. Biophys. Acta*, vol. 56, 1962, 33–42.

Garton, C. C., and Z. Krasucki, "Bubbles in insulation liquids: Stability in an electric field," *Proc. Royal Soc. (London)*, vol. A280, 1964, 211–226.

Gascoyne, P. R. C., Y. Huang, R. Pethig, J. Vykoukal, and F. F. Becker, "Dielectrophoretic separation of mammalian cells studied by computerized image analysis," *Meas. Sci. Technol.*, vol. 3, 1992, 439–445.

Gimsa, J., R. Glaser, and G. Fuhr, "Theory and application of the rotation of biological cells in rotating electric fields (electrorotation)," in *Physical Characterization of Biological Cells* (eds., W. Schütt, H. Klinkmann, I. Lamprecht, and T. Wilson). Berlin: Verlag Gesundheit, 1991, 295–323.

Glaser, R., G. Fuhr, and J. Gimsa, "Rotation of erythrocytes, plant cells, and protoplasts in an outside rotating electric field," *Studia Biophys.*, vol. 96, 1983, 11–20.

Godin, Yu. A., and A. S. Zil'bergleit, "Axisymmetric electrostatic problem of a dielectric sphere near a conducting plane," *Sov. Phys. Tech. Phys.*, vol. 31, 1986, 632–637.

Goel, N. S., and P. R. Spencer, "Toner particle-photoreceptor adhesion," in *Adhesion Science Technol.* (L. H. Lee, ed.), vol. 9, 1975, 763–829.

Goossens, K., and L. van Biesen, "Electrophoretic effects in ac dielectrophoretic separations," *Sep. Sci. Technol.*, vol. 24, 1989, 51–62.

Grebe, O., "Die Magnetpulver-Kupplung," *Elektrotech. Z.*, vol. 9, 1952, 281–284.

Griffin, J. L., "Orientation of human and avian erythrocytes in radiofrequency fields," *Exp. Cell Res.*, vol. 61, 1970, 113–120.

Gruzdev, A. D., "Orientation of microscopic particles in electric fields," *Biophysics*, vol. 10, 1965, 1206–1208.

Hagedorn, R., G. Fuhr, T. Müller, and J. Gimsa, "Traveling-wave dielectrophoresis of microparticles," *Electrophoresis*, vol. 13, 1992, 49–54.

Halsey, T. C., "Electrorheological fluids," *Science*, vol. 258, 1992, 761–766.

Halsey, T. C., and J. E. Martin, "Electrorheological fluids," *Sci. Amer.*, vol. 269, 1993, 58–64.

Halsey, T. C., J. E. Martin, and D. Adolf, "Rheology of electrorheological fluids," *Phys. Rev. Lett.*, vol. 68, 1992, 1519–1522.

Harpavat, G., "Magnetostatic force on a chain of spherical beads in a nonuniform magnetic field," *IEEE Trans. Mag.*, vol. MAG-10, 1974, 919–922.

Hartmann, G. C., L. M. Marks, and C. C. Yang, "Physical models for photoactive pigment electrophotography," *J. Appl. Phys.*, vol. 47, 1976, 5409–5420.

Hays, D. A., "Electric field detachment of toner," *Photographic Sci. Engrg.*, vol. 22, 1978, 232–235.

Hays, D. A., and W. H. Wayman, "Adhesion of a non-uniformly charged dielectric sphere," *J. Imaging Sci.*, vol. 33, 1989, 160–165.

Hebard, A. F., "A superconducting suspension with variable restoring force and low damping," *Rev. Sci. Instrum.*, vol. 44, 1973, 425–429.

Hildebrand, F. B., *Advanced calculus for applications*, Englewood Cliffs, N. J.: Prentice-Hall, 1962, Sec. 6. 14.

Holmes, L. M., "Stability of magnetic levitation," *J. Appl. Phys.*, vol. 49, 1978, 3102–3109.

Holzapfel, C., J. Vienken, and U. Zimmermann, "Rotation of cells in an alternating electric field: theory and experimental proof," *J. Membrane Biol.*, vol. 67, 1982, 13–26.

Huang, Y., R. Hölzel, R. Pethig, and X-B. Wang, "Differences in the AC electrodynamics of viable and non-viable yeast cells determined through combined dielectrophoresis and electrorotation studies," *Phys. Med. Biol.*, vol. 37, 1992, 1499–1517.

Huang, Y., and R. Pethig, "Electrode design for negative dielectrophoresis," *Meas. Sci. Technol.*, vol. 2, 1991, 1142–1146.

Inoue, A., S. Maniwa, and T. Satoh, "Electrorheological effect of modified conductive particle/dielectric fluid system," *J. Soc. Rheology (Japan)*, vol. 20, 1992, 67–72. (in Japanese).

Itoh, T., S. Masuda, and F. Gomi, "Electrostatic orientation of ceramic short fibers in liquid," *J. Electrostatics*, vol. 32, 1994, 71–89.

Jeffery, D. J., and Y. Onishi, "Electrostatics of two unequal adhering spheres," *J. Phys A: Math. Gen.*, vol. 13, 1980, 2847–2851.

Johansen, T. H., and H. Bratsberg, "Theory for lateral stability and magnetic stiffness in a high-T_C superconductor-magnet levitation system," *J. Appl. Phys.*, vol. 74, 1993, 4060–4065.

Johnson, T. W., and J. R. Melcher, "Electromechanics of electrofluidized beds," *Ind. Eng. Chem. Fundam.*, vol. 14, 1975, 146–153.

Jones, T. B., "Conditions for magnetic levitation," *J. Appl Phys.*, vol. 50, 1979a, 5057–5058.

Jones, T. B., "Dielectrophoretic force calculation," *J. Electrostatics*, vol. 6, 1979b, 69–82.

Jones, T. B., "Cusped electrostatic fields for dielectrophoretic levitation," *J. Electrostatics*, vol. 11, 1981, 85–95.

Jones, T. B., "Quincke rotation of spheres," *IEEE Trans. IAS*, vol. IA-20, 1984, 845–849.

Jones, T. B., "Multipole corrections to dielectrophoretic force," *IEEE Trans. IAS*, vol. IA-21, 1985, 930–934.

Jones, T. B., "Dielectrophoretic force in axisymmetric fields," *J. Electrostatics*, vol. 18, 1986a, 55–62.

Jones, T. B., "Dipole moments of conducting particle chains," *J. Appl. Phys.*, vol. 60, 1986b, 2226–2230 (Addendum, vol. 61, 1987, 2416–2417).

Jones, T. B., "Effective dipole moment of intersecting conducting spheres," *J. Appl. Phys.*, vol. 62, 1987, 362–365.

Jones, T. B., "Visualization of magnetic developer flows," presented at 4th Int'l. Congress on Advances in Non-Impact Printing Technologies, New Orleans, La., March 1988.

Jones, T. B., "Frequency-dependent orientation of isolated particle chains," *J. Electrostatics*, vol. 25, 1990, 231–244.

Jones, T. B., and G. W. Bliss, "Bubble dielectrophoresis," *J. Appl. Phys.*, vol. 48, 1977, 1412–1417.

Jones, T. B., and K. V. I. S. Kaler, "Relationship of rotational and dielectrophoretic cell spectra," *Proc. of IEEE/EMBS Annual Meeting,* Philadelphia, Pa., 1990, 1515–1516.

Jones, T. B., and G. A. Kallio, "Dielectrophoretic levitation of spheres and shells," *J. Electrostatics*, vol. 6, 1979, 207–224.

Jones, T. B., and J. P. Kraybill, "Active feedback-controlled dielectrophoretic levitation," *J. Appl. Phys.*, vol. 60, 1986, 1247–1252.

Jones, T. B., and L. W. Loomans, "Size effect in dielectrophoretic levitation," *J. Electrostatics*, vol. 14, 1983, 269–277.

Jones, T. B., and M. J. McCarthy, "Electrode geometries for dielectrophoretic levitation," *J. Electrostatics*, vol. 11, 1981, 71–83.

Jones, T. B., R. D. Miller, K. S. Robinson, and W. Y. Fowlkes, "Multipolar interactions of dielectric spheres," *J. Electrostatics*, vol. 22, 1989, 231–244.

Jones, T. B., M. H. Morgan, and P. W. Dietz, "Magnetic field coupled spouted bed system," U. S. Patent #4,349,967, September 21, 1982.

Jones, T. B., and D. Rubin, "Forces and torques on conducting particle chains," *J. Electrostatics*, vol. 21, 1988, 121–134.

Jones, T. B., and B. Saha, "Non-linear interactions of particles in chains," *J. Appl. Phys.*, vol. 68, 1990, 404–410.

Jones, T. B., and M. Washizu, "Equilibrium and dynamics of DEP-levitated particles," *J. Electrostatics*, vol. 33, 1994, 199–212.

Jones, T. B., G. L. Whittaker, and T. J. Sulenski, "Mechanics of magnetic powders," *Powder Technol.*, vol. 49, 1987, 149–164.

Jones, W. P., "Investigation of magnetic mixtures for clutch applications," *AIEE Trans.* (part 3), vol. 72, 1953, 88–92.

Kaler, K. V. I. S., and T. B. Jones, "Dielectrophoretic spectra of single cells determined by feedback controlled levitation," *Biophys. J.,* vol. 57, 1990, 173–182.

Kaler, K. V. I. S., and H. A. Pohl, "Dynamic dielectrophoretic levitation of living individual cells," *IEEE Trans. IAS*, vol. IA-19, 1983, 1089–1093.

Kaler, K. V. I. S., J. Xie, T. B. Jones, and R. Paul, "Dual-frequency dielectrophoretic levitation of Canola protoplasts," *Biophys. J.,* vol. 63, 1992, 58–69.

Kallio, G. A., and T. B. Jones, "Dielectric constant measurements using dielectrophoretic levitation," *IEEE Trans. IAS*, vol. IA-16, 1980, 69–75.

Kawai, H., and M. Marutake, "The dispersion of the dielectric constant in Rochelle salt crystal at low frequencies," *J. Phys. Soc. Japan*, vol. 3, 1948, 8–12.

Kendall, B. R. F., M. F. Vollero, and L. D. Hinkle, "Passive levitation of small particles in vacuum: Possible applications to vacuum gauging," *J. Vac. Sci. Tech. A*, vol. 5, 1987, 2458–2462.

Khalafalla, S. E., "Magnetic separation of the second kind: Magnetogravimetric, magnetohydrostatic, and magnetohydrodynamic separation," *IEEE Trans. Mag.*, vol. MAG-12, 1976, 455–462.

Klass, D. L., and T. W. Martinek, "Electroviscous fluids II. Electrical properties," *J. Appl. Phys.*, vol. 38, 1967, 75–80.

Klinkenberg, D. J., and C. F. Zukoski, "Studies on the steady-state behavior of electrorheological suspensions," *Langmuir*, vol. 6, 1990, 15–24.

Krasny-Ergen, W., "Nicht-thermische Wirkungen elektrischer Schwingungen auf Kolloide," *Hochfrequenztech. und Elektroakustik*, vol. 48, 1936, 126–133.

Lamb, H., *Hydrodynamics*. New York: Dover Press, 1945.

Lampa, A. S., "Dielectric hysteresis," *Sitzberichte d. k. Akademie Wissenshaft*, vol. 115, 1906, 1659–1690.

Landau, L. D., and E. M. Lifshitz, *Electrodynamics of continuous media*. Oxford: Pergamon Press, 1960.

Lebedev, N. N., and I. P. Skal'skaya, "Force acting on a conducting sphere in the field of a parallel plate condenser," *Soviet Phys.-Tech. Phys.*, vol. 7, 1962, 268–270.

Lee, M. H., "Toner adhesion in electrophotographic printers," *Proc. S. I. D.*, vol. 27, 1986, 9–14.

Lee, M. H., and J. Ayala, "Adhesion of toner to photoconductor," *J. Imaging Technol.*, vol. 11, 1985, 279–284.

Lertes, P. "Untersuchungen über Rotationen von dielectrischen Flüssigkeiten im elektrostatischen Drehfeld," *Z. Phys.*, vol. 4, 1920, 315–336.

Lertes, P. "Der Dipolrotationseffekt bei dielectrischen Flüssigkeiten," *Z. Phys.*, vol. 6, 1921, 56–68.

Levich, V. G., *Physicochemical Hydrodynamics*, Englewood Cliffs, N. J.: Prentice-Hall, 1962, chap. VIII.

Leyh, C., and R. C. Ritter, "New viscosity measurement: The oscillating magnetically suspended sphere," *Rev. Sci. Instrum.*, vol. 55, 1984, 570–577.

Liebesny, P., "Athermic short wave therapy," *Arch. Phys. Ther.*, vol. 19, 1939, 736–740.

Lin, I. J., and T. B. Jones, "General conditions for dielectrophoretic and magnetohydrostatic levitation," *J. Electrostatics*, vol. 15, 1984, 53–65.

Lo, K. Y., K. S. Ho, and I. I. Inculet, "A novel technique of electrical strengthening of soft sensitive clays," *Can. Geotech. J.*, vol. 29, 1992, 599–608.

Lorrain, P., and D. Corson, *Electromagnetic fields and waves*, 2nd ed. San Francisco, Calif.: W. H. Freeman and Company, 1970, 146–148.

Love, J. D., "The dielectric sphere-sphere problem in electrostatics," *Quart. J. Mech. Appl. Math.*, vol. 28, 1975, 449–471.

Margolis, R., unpublished work, 1988.

Marszalek, P., J. J. Zielinski, and M. Fikus, "Experimental verification of a theoretical treatment of the mechanism of dielectrophoresis," *Bioelectrochem. Bioenergetics*, vol. 22, 1989, 289–298.

Martin, C., M. Ghardiri, and U. Tüzün, "An electromechanical valve for flow control of granular materials," Proc. 2nd World Congress Particle Technology, Kyoto, Japan, September 1990, 182–191.

Masuda, S., and Itoh, T., "Electrostatic means for fabrication of fiber-reinforced metals," *IEEE Trans. IAS*, vol. MAG-25, 1989, 552–557.

Masuda, S., M. Washizu, and Nanba, T., "Novel method of cell fusion in field construction area in fluid integrated circuit," *IEEE Trans. IAS*, vol. 25, 1989, 732–737.

Maxwell, J. C., *A Treatise on Electricity and Magnetism*. New York: Dover Press, 1954, art. 314.

McAllister, I. W., "The axial dipole moment of two intersecting spheres of equal radii," *J. Appl. Phys.*, vol. 63, 1988, 2158–2160.

McLean, K. J., "Cohesion of precipitated dust layer in electrostatic precipitator," *J. Air. Pollut. Contr. Assoc.*, vol. 27, 1977, 1100–1103.

Meixner, J., "Relativistic thermodynamics of irreversible processes in a one component fluid in the presence of electromagnetic fields," University of Michigan (Radiation Laboratory) report #RL-184, April 1961.

Melcher, J. R., and G. I. Taylor, "Electrohydrodynamics: a review of the role of interfacial shear stress," *Annu. Rev. Fluid Mech.,* vol. 1, 1969, 111–146.

Meyer, R., "Nearest-neighbor approximation for the dipole moment of conducting–particle chains," *J. Electrostatics,* vol. 33, 1994, 133–146.

Miller, R. D., "Frequency-dependent orientation of lossy dielectric ellipsoids in AC electric fields," Ph. D. dissertation, Univ. of Rochester. Rochester, N. Y., 1989.

Miller, R. D., and T. B. Jones, "Frequency-dependent orientation of ellipsoidal particles in AC electric fields," Proc. of IEEE/EMBS Annual Meeting. Boston, Mass., 1987, 710–711.

Miller, R. D., and T. B. Jones, "On the effective dielectric constant of columns or layers of dielectric spheres," *J. Phys. D: Appl. Phys.,* vol. 21, 1988, 527–532.

Miller, R. D., and T. B. Jones, "Electro-orientation of ellipsoidal erythrocytes: Theory and experiment," *Biophys. J.,* vol. 64, 1993, 1588–1595.

Mischel, M., F. Rouge, I. Lamprecht, C. Aubert, and G. Prota, "Dielectrophoresis of malignant human melanocytes," *Arch. Dermatol. Res.,* vol. 275, 1983, 141–143.

Mishima, K., and T. Morimoto, "Electric field-induced orientation of myelin figures of phosphatidylcholine," *Biochim. Biophys. Acta,* vol. 985, 1989, 351–354.

Mognaschi, E. R., and A. Savini, "The action of a non-uniform electric field upon lossy dielectric systems – ponderomotive force on a dielectric sphere in the field of a point charge," *J. Phys. D: Appl. Phys.,* vol. 16, 1983, 1533–1541.

Molinari, G., and A. Viviani, "Analytical evaluation of the electro-dielectrophoretic forces acting on spherical impurity particles in dielectric fluids," *J. Electrostatics,* vol. 5, 1978, 343–354.

Moslehi, B., "Electromechanics and electrical breakdown of particulate layers," Ph. D. dissertation, Stanford Univ., Stanford, Calif., 1983.

Muth, E., "Über die Erscheinung der Perlschnurkettenbildung von Emulsionspartikelchen unter Einwirkung eines Wechselfeldes, *Kolloid-Z.,* vol. XLI, 1927, 97–102.

Oberteuffer, J. A., "High Gradient Magnetic Separation," *IEEE Trans. Mag.,* vol. MAG-9, 1973, 303–306.

Oberteuffer, J. A., "Magnetic separation: A review of principles, devices, and applications," *IEEE Trans. Mag.,* vol. MAG-10, 1974, 223–238.

Ogawa, T., "Measurement of the electrical conductivity and dielectric constant without contacting electrodes," *J. Appl. Physics,* vol. 32, 1961, 583–592.

Paranjpe, R. S., and H. G. Elrod, "Stability of chains of permeable spherical beads in an applied magnetic field," *J. Appl. Phys.,* vol. 60, 1986, 418–422.

Parmar, J. S., and A. K. Jalaluddin, "Nucleation in superheated liquids due to electric fields," *J. Phys. D: Appl. Phys.,* vol. 6, 1973, 1287–1294.

Pastushenko, V. Ph., P. I. Kuzmin, and Yu. A. Chizmadzhev, "Dielectrophoresis and electrorotation of cells: Unified theory of spherically symmetric cells with arbitrary structure of membrane," *Biological Membranes,* vol. 5, 1985, 65–77.

Pauly, H., and H. P. Schwan, "The impedance of a suspension of spherical particles surrounded by a shell," *Z. Naturforsch.,* vol. 14b, 1959, 125–131,

Pickard, W. F., "On the Born-Lertes rotational effect," *Il Nuovo Cimento,* vol. 21, 1961, 316–332.

Pohl, H. A., "The motion and precipitation of suspensoids in divergent electric fields," *J. Appl. Phys.,* vol. 22, 1951, 869–871.

Pohl, H. A., *Dielectrophoresis.* Cambridge (UK): Cambridge University Press, 1978.

Pohl, H. A., and J. S. Crane, "Dielectrophoresis of cells," *Biophys. J.,* vol. 11, 1971, 711–727.

Price, J. A. R., J. P. H. Burt, and R. Pethig, "Applications of a new optical technique for measuring the dielectrophoretic behaviour of micro-organisms," *Biochim. Biophys. Acta*, vol. 964, 1988, 221–230.

Quincke, G., "Electrische Untersuchungen: XIV. Über Rotationen im konstanten electrischen Felde," *Ann. Phys. Chem.*, vol. 59, 1896, 417–486.

Ramakrishnan, S., and K. P. P. Pillai, "Theory and performance of the disc-type electromagnetic particle clutch under continuous slip service," *IEE Proc.*, vol. 127, 1980, 81–88.

Richards, A. H., J. G. Magondu, R. N. W. Laithwaite, and P. N. Murgatroyd, "Self-generated rotation in a magnetic levitator," *IEE Proc. (London)*, vol. 128, 1981, 449–452.

Robinson, K. S., "Electromechanics of packed granular beds," Ph. D. dissertation, Colo. State. Univ., Ft. Collins, Colo., 1982.

Robinson, K. S., and T. B. Jones, "Particle-wall adhesion in electropacked beds," *IEEE Trans. Ind. Applic.*, vol. IA-20, 1984, 1573–1577.

Rosensweig, R. E., "Fluidmagnetic buoyancy," *AIAA J.*, vol. 4, 1966, 1751–1758.

Rosensweig, R. E., "Fluid dynamics and science of magnetic liquids," *Adv. Electronics Electron Phys.*, vol. 48, 1979a, 103–199.

Rosensweig, R. E., "Fluidization: Hydrodynamic stabilization with a magnetic field," *Science*, vol. 204, 1979b, 57–60.

Rosensweig, R. E., G. R. Jerauld, and M. Zahn, "Structure of magnetically stabilized fluidized solids," in *Continuum Models of Discrete Systems*, O. Brulin and R. K. T. Hsieh, eds., Amsterdam: North-Holland Publishing Co., 1981, 137–144.

Saito, M., and H. P. Schwan, "The time constants of pearl-chain formation," in *Biological Effects of Microwave Radiation*, vol. I. N. Y.: Plenum Press, 1961, 85–97.

Saito, M., G. Schwarz, and H. P. Schwan, "Response of nonspherical biological particles to alternating electric fields," *Biophys. J.*, vol. 6, 1966, 313–327.

Sauer, F. A., "Forces on suspended particles in the electromagnetic field," in *Coherent excitations in biological systems*, H. Fröhlich and F. Kremer (eds.), Berlin: Springer-Verlag, 1983, 134–144.

Sauer, F. A., "Interaction forces between microscopic particles in an external electromagnetic field," in *Interactions between electromagnetic fields and cells*, A. Chiabrera, C. Nicolini, and H. P. Schwan (eds.). New York: Plenum, 1985, 181–202.

Sauer, F. A., and R. W. Schlögl, "Torques exerted on cylinders and spheres by external electromagnetic fields. A contribution to the theory of field induced rotation" in *Interactions between electromagnetic fields and cells*, A. Chiabrera, C. Nicolini, and H. P. Schwan (eds.). New York: Plenum, 1985, 203–251.

Schwan, H. P., and L. D. Sher, "Alternating-current field-induced forces and their biological implications," *J. Electrochem. Soc: Reviews and News*, vol. 116, 1969, 22C–26C.

Schwarz, G., "A Theory of the low-frequency dielectric dispersion of colloidal particles in electrolyte solution," *J. Phys. Chem.*, vol. 66, 1962, 2636–2642.

Schwarz, G., M. Saito, and H. P. Schwan, "On the orientation of nonspherical particles in an alternating electric field," *J. Chem. Phys.*, vol. 43, 1965, 3562–3569.

Scott, D., W. W. Seifert, and V. C. Westcott, "The particles of wear," *Sci. Amer.*, vol. 230, 1974, 88–97.

Secker, P. E., and I. N. Scialom, "A simple liquid-immersed dielectric motor," *J. Appl. Phys.*, vol. 39, 1968, 2957–2961.

Shalom, A. L., and I. J. Lin, "Frequency dependence on dielectrophoresis: Trajectories and particle buildup," *Chem. Engrg. Comm.*, vol. 74, 1988, 111–121.

Sillars, R. W., "The properties of a dielectric containing semiconducting particles of various shapes," *J. Inst. Elect. Eng.* (London), vol. 80, 1937, 378–394.

Simpson, P., and R. J. Taylor, "Characteristic rotor speed variations of a dielectric motor with a low-conductivity liquid," *J. Phys. D: Appl. Phys.,* vol. 4, 1971, 1893–1897.

Smythe, W. R., *Static and dynamic electricity.* New York: McGraw-Hill, 1968, Sec. 10. 06.

Stenger, D. A., and S. W. Hui, "Human erythrocyte electrofusion kinetics monitored by aqueous contents mixing," *Biophys. J.,* vol. 53, 1988, 833–838.

Stenger, D. A., K. V. I. S. Kaler, and S. W. Hui, "Dipole interactions in electrofusion," *Biophys. J.,* vol. 59, 1991, 1074–1084.

Stepin, L. D., "Dielectric permeability of a medium with nonuniform ellipsoidal inclusions," *Soviet Phys. Tech. Phys.,* vol. 10, 1965, 768–772.

Stoy, R. D., "Solution procedure for the Laplace equation in bispherical coordinates for two spheres in a uniform external field: Parallel orientation," *J. Appl. Phys.,* vol. 65, 1989a, 2611–2615.

Stoy, R. D., "Solution procedure for the Laplace equation in bispherical coordinates for two spheres in a uniform external field: Perpendicular orientation," *J. Appl. Phys.,* vol. 66, 1989b, 5093–5095.

Stoylov, S. P., *Colloid electro-optics: Theory, techniques, applications.* London: Academic Press, 1991.

Stratton, J. A., *Electromagnetic theory,* New York: McGraw-Hill Book Company, 1941.

Sumoto, I., "An interesting phenomenon observed on some dielectrics," *J. Phys. Soc. Japan,* vol. 10, 1955, 494.

Sumoto, I., "Rotary motion of dielectrics in a static field," *Rept. Sci. Res. Inst. (Tokyo),* vol. 32, 1956, 41–46 (in Japanese).

Talbert, C. M., T. B. Jones, and P. W. Dietz, "The electro-spouted bed," *IEEE Trans. Ind. Applic.,* vol. IA-20, 1984, 1220–1223.

Talbott, J. W., and E. K. Stefanakos, "Aligning forces on wood particles in an electric field," *Wood and Fiber,* vol. 4, 1972, 193–203.

Tan, C. unpublished data, 1993.

Tan, C., and T. B. Jones, "Interparticle force measurements on ferromagnetic steel balls," *J. Appl. Phys.,* vol. 73, 1993, 3593–3598.

Teissie, J., V. P. Knudson, T. Y. Tsong, and M. D. Lane, "Electric pulse-induced fusion of 3T3 cells in monolayer culture," *Science,* vol. 216, 1982, 537–538.

Tombs, T. N., and T. B. Jones, "Digital dielectrophoretic levitation," *Rev. Sci. Instrum.,* vol. 62, 1991, 1072–1077.

Tombs, T. N., and T. B. Jones, "Effect of moisture on the dielectrophoretic spectra of glass spheres," *IEEE Trans. IAS,* vol. 29, 1993, 281–285.

Torza, S., R. G. Cox, and S. G. Mason, "EHD deformation and burst of liquid drops," *Phil. Trans Royal Soc. (London),* vol. A269, 1971, 295–319.

Turcu, I., "Electric field induced rotation of spheres," *J. Phys. A: Math. Gen.,* vol. 20, 1987, 3301–3307.

Veas, F., and M. J. Schaffer, "Stable levitation of a dielectric liquid in a multiple-frequency electric field," Proc. of International Symposium on Electrohydrodynamics, Mass. Inst. of Technol., March–April 1969, 113–115.

Vienken, J., U. Zimmermann, A. Alonso, and D. Chapman, "Orientation of sickle red blood cells in an alternating electric field," *Naturwissenschaften,* vol. 71, 1984, 158–160.

von Guggenberg, P. A., A. J. Porter, and J. R. Melcher, "Field-mediated hydraulic deformation and transport of magnetically solidified magnetizable particles," *IEEE Trans. Mag.,* vol. MAG-22, 1986, 614–619.

Wang, X-B., R. Pethig, and T. B. Jones, "Relationship of dielectrophoretic and electrorotational behavior exhibited by polarized particles," *J. Phys. D: Appl. Phys.,* vol. 25, 1992, 905–912.

Washizu, M. "Electrostatic manipulation of biological objects," *J. Electrostatics*, vol. 25, 1990, 109–123.

Washizu, M., and T. B. Jones, "Multipolar dielectrophoretic force calculation," *J. Electrostatics*, vol. 33, 1994, 187–198.

Washizu, M., T. B. Jones, and K. V. I. S. Kaler, "Higher-order DEP effects: Levitation at a field null," *Biochim. Biophys. Acta*, vol. 1158, 1993a, 40–46.

Washizu, M., Y. Kurahashi, H. Iochi, O. Kurosawa, S. Aizawa, S. Kudo, Y. Magariyama, and H. Hotani, "Dielectrophoretic measurement of bacterial motor characteristics," Proc. of IEEE/IAS Annual Meeting, Detroit, MI, 1991, 665–673.

Washizu, M., and O. Kurosawa, "Electrostatic manipulation of DNA in microfabricated structures," *IEEE Trans. IAS*, vol. 26, 1990, 1165–1172.

Washizu, M., O. Kurosawa, I. Arai, S. Suzuki, and S. Shimamoto, "Applications of electrostatic stretch and positioning of DNA," Proc. of IEEE/IAS Annual Meeting, Toronto, Canada, 1993b, 1629–1637.

Washizu, M., and K. Nakada, "Bearingless micromotor using negative dielectrophoresis," in Proceedings of *Annual Meeting of Institute of Electrostatics – Japan*, Tokyo, 1991, 233–236 (in Japanese).

Washizu, M., T. Nanba, and S. Masuda, "Handling biological cells using a fluid integrated circuit," *IEEE Trans. IAS*, vol. 26, 1990, 352–358.

Washizu, M., M. Shikida, S. Aizawa, and H. Hotani, "Orientation and transformation of flagella in electrostatic field," *IEEE Trans. IAS*, vol. 28, 1992a, 1194–1202.

Washizu, M., S. Suzuki, O. Kurosawa, T. Nishizaka, and T. Shinohara, "Molecular dielectrophoresis of bio-polymers," *Proc. of IEEE/IAS Annual Meeting,* Houston, Tex., 1992b, 1446–1452.

Watson, J. H. P., "Magnetic filtration," *J. Appl. Phys.,* vol. 44, 1973, 4209–4213.

Weber, E., *Electromagnetic theory.* New York: Dover, 1965, sec. 21.

Weiler, W., "Zur Darstellung elektrischer Kraftlinien," *Z. phys. chem. Unterricht*, Heft IV, April 1893, 194–195.

Williams, T. J., private communication, 1993.

Winslow, W. M., "Induced fibration of suspensions." *J. Appl. Phys.,* vol. 20, 1949, 1137–1140.

Woodson, H. H., and J. R. Melcher, *Electromechanical Dynamics,* New York: Wiley, 1968.

Yang, W., E. Jaraiz M., O. Levenspiel, and T. J. Fitzgerald, "A magnetic control valve for flowing solids: Exploratory studies," *Ind. Eng. Chem. Process Des. Dev.,* vol. 21, 1982, 717–721.

Yarmchuk, E. J., and J. F. Janak, "Chains of permeable spheres in an applied magnetic field," *IEEE Trans. Mag.,* vol. MAG-18, 1982, 1268–1270.

Zahn, M., and Shi-Woo Rhee, "Electric field effects on the equilibrium and small signal stabilization of electrofluidized beds, *IEEE Trans. IAS*, vol. IA-20, 1984, 137–147.

Zimmermann, U., and J. Vienken, "Electric field induced cell to cell fusion," *J. Membrane Biol.,* vol. 67, 1982, 165–182.

Zimmermann, U., J. Vienken, and G. Pilwat, "Electrofusion of cells," in *Investigative Microtechniques in Medicine and Biology,* J. Chayen and L. Bitensky (eds.). New York: Marcel Dekker, 1984, 89–167.

Index

acceleration, gravitational, 51
adhesion, electrostatic contribution to, 2, 194
admittance, series and shunt, 20, 230
alignment. *See* orientation
analogy, electric to magnetic, 6, 34, 63, 132, 142, 218
Andres, U., 63
angle, lag, 85, 101, 104
anisotropy
   material, 68, 110, 127, 177, 216
   shape-dependent, 110, 122, 153
Argand diagram, 21, 48, 98, 240
arrays, particle, 189, 209
axis, semimajor, 111

bed, particle
   electrofluidized, 2, 214, 220
   electrospouted, 2, 215, 220
   electrostabilized, 216
   magnetopacked, 215
   magnetospouted, 220
   magnetostabilized, 2, 141, 215
Bessel functions, 70
Boltzmann constant, 141
Born–Lertes effect. *See also* electrorotation, 83
boundary conditions, 11, 17, 219, 228, 231, 233
Brownian motion, 3, 76
brush, magnetic. *See also* tumbling, 2, 141
bubble theory of breakdown, 111

capacitance
   membrane, 48, 77, 232
   surface, 21, 124
cell, biological, 43, 74, 105, 110, 121, 137, 211, 232, 246
chain, particle

conducting, 140
dielectric, 159
magnetic, 168, 189, 191, 209
orientation of, 212
orientation of, frequency-dependent, 172
charge, point source, 14, 142, 147
Clausius–Mossotti function, 11, 13, 16, 18, 24, 29, 39, 65, 70, 85, 102, 139, 220, 228, 238
clutch, magnetic, 216
cohesion
   electric, 214
   magnetic, 210
Cole–Cole plot, 244
conductance
   surface, 233
   transmembrane, 232, 246
conductivity, generalized, 218
Coulomb's Law, 15, 79, 143, 182
current, surface, 21, 233

damping, viscous, 53
DEP. *See* dielectrophoresis
depolarization, 113, 118, 161
dielectrophoresis (DEP), 34
   approximation, 8, 62
   biological, 44, 74
   negative, 37, 43, 52, 55
   positive, 37, 43, 52, 58
dipole
   approximation, 161, 162, 188, 209, 250
   effective, method of, 25, 217
   finite, 222
   force acting on, 7, 24, 28
   induced, 9, 161
   infinitesimal, 6
   permanent, 135, 207
   torque acting on, 8, 24, 28, 29, 30, 86

DNA, 137

eddy current, 63, 69, 101, 132
electrofusion, 75, 105, 211
electrohydrodynamics (EHD), 36
electromechanics, 2, 141, 181, 189, 192,
    194, 208, 216
electrophoresis, 2, 36
electrorotation, 83, 92, 106
ellipsoid. *See also* spheroid, 61, 111, 132,
    177, 251
equilibrium, rotational, 87, 91, 100
expansion
    multipolar, 15, 223, 234
    field, Taylor-series, 53

ferrofluid, 67, 68
ferrography, 81
field
    electric, axisymmetric, expansion of, 53
    electric, maxima and minima, 50
    electric, rotating, 30, 84, 90, 128
    magnetic, maxima and minima, 66
    magnetic, rotating, 101, 135
fluid, electrorheological, 1, 34, 207, 212
flux, magnetic, trapped, 74
force
    coercive, magnetic, 193
    current-controlled interparticle, 187,
        215
    detachment, 203, 207
    dielectrophoretic, 26, 39, 118
    interparticle, 182, 184, 215
    lifting, 184
    magnetophoretic, 65
    ponderomotive, 6
Fröhlich, H., 63

gain, feedback, 59
Gauss's Law, 18

hardness, magnetic, 67
Harpavat, G., 141, 160, 188, 209
Helmholtz coil, 188
hysteresis, magnetic, 67, 104

image
    force, 196, 207
    method of, 142, 145, 183
    theory, 159, 166
inertia, moment of, 99
interactions
    imposed field, 5

mutual particle, 5, 139, 160, 173, 180,
    197, 207
inversion, geometric, method of, 142, 149

Krämers–Kronig relations, 94, 243

laminae, 135
Laplace's equation, 10, 43
Legendre polynomials, 15, 223
Levich, V. G., 36
levitation equilibria
    field null, 56
    marginal radial stability, 56
    release point, 56
    stability, 55, 59
levitation
    dielectrophoretic, 40, 49, 77, 156, 250
    feedback-controlled, 52, 58, 66, 77
    magnetic, 66, 72, 81
    method of fixed position, 60
    passive DEP, 52, 55, 60, 100
Lorenz force law, 26, 220
loss
    dielectric, 18, 28, 31, 33
    ohmic, 16, 18, 20, 28, 233

magnetism, nonlinear, 66, 103, 192
magnetization, remanent, 63, 105, 132,
    135, 193
magnetofluid buoyancy, 68
magnetometer, vibrating sample, 168
magnetophoresis (MAP), 34, 62, 70
    negative, 66, 73
    positive, 66, 73
mass
    buoyant, 51
    density, 53
    effective, 53
Maxwell
    charge, 144, 185
    equations, 220
    mixture theory, 175, 229
    stress tensor, 9, 24, 28, 30, 185
Maxwell–Wagner relaxation, 19, 20, 42,
    84, 94, 105, 119, 168, 230, 236, 239
Meissner effect, 73
microactuators, 78
mixture, heterogeneous, 173
moment
    dipole, 7, 24, 138, 203
    dipole, rotating, 85, 90, 101
    electric, effective, 9, 13, 18, 24, 31, 84,
        105, 112, 144, 153, 156, 161, 172,
        237, 252

magnetic, effective, 64, 70, 101, 104, 133, 192
multipolar, 159, 168, 189, 235, 249
octupolar, 166
quadrupolar, 147, 161, 207, 250
multipole
  force acting on, 29, 225, 249
  general, 27, 248
  higher order, 38, 224
  linear, 15, 26, 164, 181, 182, 222, 224

Newton's first law, 53
orientation
  parallel, 115, 118, 134, 150, 170
  particle, 110, 141
  perpendicular, 115, 134, 151, 170

particle
  bed, 209, 214
  chain, 140, 145, 147, 181, 207, 211
  crystalline, 127, 133, 134, 136
  diamagnetic, 63, 72, 73
  dielectric, 136
  ellipsoidal, 111, 118, 251
  ferromagnetic, 63, 67, 136, 168
  layered, 13, 42, 121, 227
  lossless, 12, 25, 227
  magnetizable, 63, 132
  paramagnetic, 63
  superconducting, 73, 74
patch charge model, 199
pearl chains, 140
permittivity, complex. *See also* Cole–Cole
  plot, 17, 119
Pohl, H. A., 3, 35, 84
polarization
  excess, 112, 239
  Maxwell–Wagner. *See* Maxwell–Wag-
    ner relaxation
pole shading, 106
potential
  electromechanical, 50
  electrostatic, 9, 14, 20, 25, 150, 219, 221, 226, 251
  magnetic vector, 69
  magnetostatic, 63, 219, 221
protoplast, 46, 96, 246

quadrupole, 147, 161, 207, 250
Quincke, G., 84, 98, 110

relaxation, charge, 94, 237
reluctance, magnetic, 193
resistance, surface, 233
Reynolds number, magnetic, 71, 102

Riemann–zeta function, 144, 145, 151
ROT. *See* electrorotation
rotation
  Quincke, 98, 108, 110
  synchronous, 104, 129

saliency
  particle, 129
  electrode, 250
saturation, magnetic, 66, 132, 170, 193
Sauer, F. A., 28, 30
separation
  dielectric, 34, 42, 79, 108
  magnetic. *See also* Andres, U., 2, 34, 80, 138
shell, spherical, 12, 20, 227, 229, 234
spectrum
  dielectrophoretic, 29, 40, 61, 77, 238
  frequency, 22
  orientational, 111, 120, 124, 173
  rotational, 92, 94, 238
sphere
  conducting, 12, 17, 221
  conducting, perfectly, 142, 150
  dielectric, 10, 86, 159, 221
spheres, intersecting, 155
spheroid
  oblate, 117, 128, 135
  prolate, 115, 129, 134
stagnant layer, 210
Stokes drag formula, 53
Stoylov, S. P., 4
superconductivity, 74
surface tension, 194
susceptibility, magnetic, 133
suspension, electric, 215

tensor
  moment, 27, 249
  permittivity, 127
  susceptibility, 133
time constant
  charge relaxation, 19, 21, 94
  thermal, 140
torque
  drag, viscous, 88
  orientational, 110, 130, 135, 153
  rotational, 86, 110, 128
transient behavior, 16, 42, 236
tumbling, magnetic. *See also* brush, magnetic, 208
Turcu, I., 84, 88

VSM. *See* magnetometer, vibrating sample

Made in the USA
Lexington, KY
29 May 2012